T0076495

Difficult Engineering Concepts Better Explained

—— Statics and Applications ——

Difficult Engineering Concepts Better Explained

——Statics and Applications——

JAY F TU

North Carolina State University, USA

World Scientific

NEW JERSEY · LONDON · SINGAPORE · BEIJING · SHANGHAI · HONG KONG · TAIPEI · CHENNAI · TOKYO

Published by

World Scientific Publishing Co. Pte. Ltd.

5 Toh Tuck Link, Singapore 596224

USA office: 27 Warren Street, Suite 401-402, Hackensack, NJ 07601

UK office: 57 Shelton Street, Covent Garden, London WC2H 9HE

Library of Congress Cataloging-in-Publication Data

Names: Tu, Jay F., author.

Title: Difficult engineering concepts better explained : statics and applications /
Jay F Tu, North Carolina State University, USA.

Description: Singapore ; Hackensack, NJ : World Scientific Publishing Co. Pte. Ltd., [2020] |
Includes bibliographical references and index.

Identifiers: LCCN 2020006590| ISBN 9789811213786 (hardcover) | ISBN 9789811213793 (ebook) |
ISBN 9789811213809 (ebook other)

Subjects: LCSH: Statics. | Mechanics, Applied.

Classification: LCC TA351 .T825 2020 | DDC 320.1/03--dc23

LC record available at https://lccn.loc.gov/2020006590

British Library Cataloguing-in-Publication Data

A catalogue record for this book is available from the British Library.

For any available supplementary material, please visit
https://www.worldscientific.com/worldscibooks/10.1142/11652#t=suppl

Desk Editors: Anthony Alexander/Steven Patt

Typeset by Stallion Press
Email: enquiries@stallionpress.com

Preface

In his Preface to the famous three volumes of *The Feynman Lectures on Physics*, the legendary physicist, Dr. Richard P. Feynman, wrote,

"In the first part of the course, dealing with electricity and magnetism, I couldn't think of any really unique or different way of doing it — of any way that would be particularly more exciting than the usual way of presenting it."

And then he hoped,

"Perhaps they (students) will have fun thinking them through — or going on to develop some of the ideas further."

These two statements by Dr. Feynman pretty much sum up the motivation of this book. Can I present undergraduate Engineering Statics in a unique or different way? Can I explain difficult concepts better or more excitingly? Can I instill students with enough motivations to think deeper and to develop the ideas further? After all, learning is not simply memorizing how to solve standardized problems. In view of many, many excellent textbooks on Statics, some at their 15th edition or higher, it is apparently a tall order.

However, if one actually had ventured into the contents of Feynman's lectures, he or she would have been wonderfully entertained by how a genius mind works. A few of Feynman's signature characteristics jumped out at me when I read his lectures. One is his ability to explain difficult concepts with vivid but common life experiences. For example, in explaining the least-time principle of light's traveling path, Feynman gave an example of how a lifeguard should rush to a drowning person on a beach in the shortest amount of time. Should the guard run straight to the person in a straight line, through the sand and the ocean water, which is the shortest distance, or run on the sand a bit more, not directly toward the person, before jumping into ocean toward the person? Because running is faster than swimming, the latter approach can achieve the shortest amount of time to reach the victim, which happens to explain the "intelligence" of light that it can choose the path with the shortest amount of time by bending its path when entering glass from air. However, this "intelligence" of light is later exposed as not true when the ideas are developed further. Light after all is not a living being to possess the intelligence of thinking.

Although far from the stratosphere of Feynman's intellectual capabilities, after over 28 years in higher education as an engineering professor, I have developed a few bags of tricks that I feel are worth sharing. Statics and Dynamics present substantial challenges for Mechanical Engineering students because they form the foundation for later more advanced subjects in material strength, fluid mechanics, manufacturing, and engineering design. With standardized test problems and the all-out attacks of smart phones and Internet search capabilities, the teaching of these two fundamental subjects, as well as many other subjects, has become more like short-term memory exercises. Students do not "have fun thinking them through" but simply recite the solutions they have practiced earlier, often just one or two nights before. Many times, I notice that students' brains would cease to function if they are confronted with problems they never practiced before. If allowed to get help, students would "Google" for answers or simply venture into some statements based on their "gut feeling" without proper scientific analysis based on basic laws. In other words, they would speculate and hope they would get lucky and be awarded some partial credits for offering speculations. We do not design airplanes or cars by speculations; if I may paraphrase Einstein's words, "God does not play dice with the universe."

We, in higher education, must not forget the importance of motivating students to think through difficult problems, while having fun doing so. It takes major effort to think, but one who does so will be rewarded generously.

I don't have the illusion that this book would change this short-term memory learning culture, but I do hope that I can reach out to a few students who want to learn these subjects and other engineering subjects in an, as Feynman put it, "exciting and interesting way." As a result, this book is not written to compete with any standard textbooks on Statics, but as a supplement for students to look at the key concepts more fundamentally and to relate them to real-world examples. Another objective is to reduce the number of equations students feel that they need to memorize. After all, the entire Statics and Dynamics are based on just one equation, $\sum \vec{F} = m\vec{a}$, from Sir Newton.

Finally, in this book, all the graphics, if not drawn by myself personally, are obtained from "bing image search" via Microsoft Word and only limited to those from "Creative Commons". All these graphics are provided with the links where they are from, to observe their copyrights.

About the Author

 **Jay Tu** graduated from National Taiwan University in 1981 and received his Ph.D. degree in Mechanical Engineering from the University of Michigan, Ann Arbor, in 1991. He is currently a full professor at North Carolina State University (NCSU). Before joining NCSU in 2003, he was an assistant professor and a tenured associate professor at Purdue University from 1992 to 2003.

Dr. Tu works closely with industry to conduct high-impact research to enhance the precision, reliability, and productivity of modern manufacturing systems through fundamental research in modeling, testing, and design. His specific research areas include high-speed machining, high-speed spindles, and laser material processing, all of which involve complex optical, electrical, and mechanical systems. Dr. Tu has infused his research interests into his teaching in both his graduate and undergraduate courses, with a strong emphasis on enhancing the thinking capabilities of students. In his classes, he presents numerous real-world examples and stories to show how fundamental laws can be used to solve practical engineering problems and to explain common events in music, sports, and daily life.

Contents

Chapter 1

Mechanics I: Principles of Statics

Principles of Statics is a subset of Mechanics that deals with bodies at rest despite being under the action of forces. The bodies are considered to be at equilibrium when all the forces sum to zero. To study Statics, we first need to understand the nature of forces, quantify them, and then understand how forces can be applied to bodies. The relevant fundamental concepts and mathematical tools related to forces are discussed in Chapters 2–4. The main objective of studying Statics is to conduct correct force analysis for mechanical or integrated systems for design, performance improvement, or failure prevention. The standard method for force analysis, called free-body diagram analysis, is presented in Chapters 5 and 6. In Chapter 7, we demonstrate how proper force analysis can be conducted for various practical applications. However, often, a system might be too complicated to be analyzed by the free-body diagram analysis. Chapter 8 addresses this situation. In Chapter 9, we discuss the stability of equilibrium. In Chapter 10, we present a case study to demonstrate how to conduct proper force analysis to design a sensor for monitoring machining forces. Finally, in Chapter 11, we provide detailed solutions to 31 difficult problems in Statics.

As stated, the main objective of studying Statics is to conduct correct force analysis for mechanical or integrated systems. To this end, we, as engineers, must approach the analysis quantitatively, with proper knowledge about precision and uncertainties. We will discuss key concepts and practices to ensure confidence in the calculations.

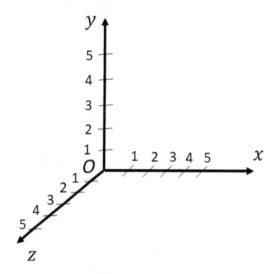

Figure 1.1:   A Cartesian coordinate with metrics defined along each axis.

Finally, in Chapter 1, we will suggest a general problem-solving protocol to promote a systematic and creative thought process. Again, the idea is that we can do a little thinking to solve new problems, not limited to the problems we have studied before.

1.1 Space and Time

Space and time (or spacetime) are fundamental metrics we need for solving problems. To quantify a space, typically, we will define a reference coordinate. Typically, in Mechanics, we deal with a 1D, 2D, or 3D space, defined commonly by a suitable coordinate such as the $O - x - y - z$ coordinate, shown in Fig. 1.1. Each axis of the $O - x - y - z$ coordinate is typically fixed in direction in Statics.

We need to define metrics along each axis. In other words, how do we define length, for example, the length of one meter, in the SI system. In the old days, there was a "National Prototype Metre Bar No. 27, made in 1889 by the international Bureau of Weights and Measures (BIPM) and it was given to the United States to serve as the standard for defining all units of length in the US from 1893 to 1960."[1] "The bars were to be made of a special alloy, 90% platinum and 10% iridium, which is significantly harder

---

[1] https://en.wikipedia.org/wiki/History_of_the_metre#International_prototype_metre.

than pure platinum, and have a special X-shaped cross section (a 'Tresca section', named after French engineer Henri Tresca) to minimize the effects of torsional strain during length comparisons."[2]

Today, the standard of length is no longer defined by a meter bar, but by time, based on Einstein's theory of special relativity, which assumes that the speed of light in vacuum is constant. The official definition is[3]

"The metre, symbol m, is the SI unit of length. It is defined by taking the fixed numerical value of the speed of light in vacuum c to be 299792458 when expressed in the unit m/s, where the second is defined in terms of the caesium frequency $\Delta\nu$Cs."

Therefore, length is based on time. The question now is how do we achieve precise time-keeping? According to NIST,[4] "since 1967, the International System of Units (SI) has defined a second as the period equal to 9,192,631,770 cycles of the radiation, which corresponds to the transition between two energy levels of the ground state of the Cesium-133 atom. This definition makes the cesium oscillator (sometimes referred to generically as an atomic clock) the primary standard for time and frequency measurements." In other words, we rely on counting to establish time and thus the standard of length.

1.2 Concept of Precision

Being quantitative is one important requirement for engineers. We need to handle numbers properly.

In Section 1.1, we talked about using counting for time-keeping, precisely, at least down to $1/9,192,631,770$ seconds, or 0.10873 nanoseconds based on the definition of 1 second.

As we count, we could make mistakes or simply reach different counts because the event has changed. In other words, there are errors and uncertainties. Errors are due to mistakes. The uncertainties are due to factors out of our control, no matter how carefully we try to avoid mistakes. The subject of precision deals with errors and uncertainties. There are

---

[2]Nelson, R.A. (1981). "Foundations of the international system of units (SI)". *The Physics Teacher.* **19**(9): 596–613.
[3]BIPM. (2019). SI Brochure, 9th edition, p. 131.
[4]https://www.nist.gov/pml/time-and-frequency-division/timekeeping-and-clocks-faqs.

three main concepts in precision, which are accuracy, repeatability, and resolution.[5]

For example, if we use a ruler to measure the length of a rod, the accuracy is the deviation of the measurement from the true value of the length. The problem is that we usually do not know the true value. When we take measurement with a ruler, the ruler has its limit of resolution, which is, for example, 1 mm, for a simple office ruler. We can get a 0.01 mm resolution using a caliper.

With a simple ruler, we align one end of the rod to the zero mark of the ruler and check where the other end ends. If the other end lands between the marks of 11 and 12 mm, we can only guess that the true length is, judging where it lands between the marks, 11.4 mm, for example. In fact, we are not sure if it is 11.3, 11.4, or 11.5 mm, and different people might guess it differently. As a result, we have an uncertainty, which is at least 0.2 mm, in this case. This uncertainty affects how repeatable the measurement could be.

One practical way of achieving a higher precision is to make multiple measurements. For example, if we want to estimate the value of an antique piece, we can ask many experts to estimate its value. Because we cannot trust one expert more than others, the best practice is to calculate the average of all the estimates. The average, thus, is the best estimate to incorporate all the inputs from all the experts. To determine the accuracy, we should use the best estimate (the average) and calculate its deviation from the true value (if we know it). The variation among the estimates from different experts represents repeatability. The resolution is related to the smallest money unit used in the estimates for this case.

1.2.1 Significance of figures

When we make a measurement or present a number, we often do not explicitly state its precision in terms of accuracy, repeatability, and resolution. Instead, we use the concept of significant figures to indicate the precision of a measurement. This is the subject of arithmetic precision. "A significant figure is any one of the digits 1 to 9; zero is a significant figure except when it is used to fix the decimal point or to fill the places of unknown or discarded digits."[6] "Thus in the number 0.000532, the

---

[5]Slocum, A. (1992). *Precision Machine Design*. Society of Manufacturing Engineers, Dearborn, Michigan. ISBN-13: 978-0872634923.

[6]Doebelin, E.O. (1990). *Measurement Systems Application and Design*. McGraw-Hill Publishing Company, New York, NY. ISBN 0-07-017338-9.

significant figures are 5, 3, and 2, while in the number 2,076, all the digits, including the zero, are significant. For a number such as 2,300, the zeros may or may not be significant. To convey which figures are significant, we write this as 2.3×10^3 if two significant figures are intended, 2.30×10^3 if three, 2.300×10^3 if four, and so forth (see footnote 6)."

We often have to compute by using many measurements with different significant figures. When we are doing hand calculation, it is preferred to round the number with more significant figures to be the same as the one with less significant figures. There are practical reasons for this rounding practice such as (1) the final results cannot have a higher number of significant figures and (2) handling a long string of digits can incur more mistakes. However, in today's computing practice using computers, the second concern is no longer an issue. We should still round the number of the final result to recognize its limited significant figures and to enable easy reading.

A widely used rounding practice is as follows (see footnote 6):

"To round a number to n significant figures, discard all digits to the right of the nth place. If the discarded number is less than one-half a unit in the nth place, leave the nth digit unchanged. If the discarded number is greater than one-half a unit in the nth place, increase the nth digit by 1. If the discarded number is exactly one-half a unit in the nth place, leave the nth digit unchanged if it is an even number and add 1 to it if it is odd."

Therefore, if we are given a number, 12.35, we should know that the actual number could range from 12.346 to 12.355, or $12.35^{+0.005}_{-0.004}$. The uncertainty range could be larger than 0.009 because most people may not follow the rounding rule consistently.

In some applications, we need to emphasize the number of decimal places (the number of significant figures following the decimal point). For example, in your bank statement, the balance of your checking account is rounded to two places after the decimal point. This is of course for practical reasons because we do not have a coin smaller than one cent.

In general, when one is performing addition and/or subtraction, it is better to round the result to the same number of decimal places as the number with the lowest number of decimal places. When the computation involves multiplication and/or division, it is better to round the result to be the same as the one with the lowest significant figures.

In this book, armed with modern computing power, all numbers of final results will be rounded to two decimal places unless a higher precision is required. If a number is very small, such as 0.0017345, it would be rounded to 0.00173, to contain three significant figures, unless a higher precision is required.

1.2.2 Statistical nature of uncertainty

In the above discussion, we specified the range of uncertainty of a measurement using the plus and minus expression. If a number is expressed as $12.35^{+0.005}_{-0.004}$, we only specify that all the measurements would be between 12.346 and 12.355. To understand this uncertainty more, we should also define the probability distribution within the range of uncertainty. Essentially, there are no fundamental laws to help us decide how often a number would appear within the range of uncertainty.

To answer this question, a basic discussion on statistics is in order. If we measure the blood pressure of a person several times, it is highly unlikely we will get the same reading. Let us say that we take the measurement 100 times, consecutively (hypothetically, not advised to do so), and obtained the measurements of the systolic pressure as shown in Fig. 1.2 (this figure is denoted as a histogram).

The measurements are grouped into 26 bins and each bin covers 2 mmHg. From Fig. 1.2, we found that, for example, there are 15 readings between 106 and 108 mmHg, only 2 readings over 130 mmHg, 10 readings between 120 and 130 mmHg, and 5 readings below 90 mmHg. Therefore, is the blood pressure of the person normal? There are 83 readings within the

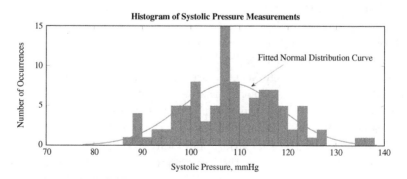

Figure 1.2:   Histogram of 100 measurements of systolic pressure.

normal range between 90 and 120 mmHg, but 12 times when the pressure is in the pre–high blood pressure range. What is causing all these variations? The actual blood pressure could fluctuate, and the measurement device could have measurement errors. The question is still "Is it normal?" The same scenario can be related to the strength of a structure or the quality of a part after we obtain multiple measurements.

One common statistical practice is to assume that the probability distribution of the reading is a normal distribution. We can fit a normal distribution curve to the histogram, as shown in Fig. 1.2. This assumption is typically valid due to the Central Limit Theorem, or we just assume that it is for convenience. There are, of course, many exceptions. Based on the fitted normal distribution, we assume that if we repeat the measurement an almost infinite number of times, the histogram will approach a normal distribution curve (not necessary the same as the fitted one in Fig. 1.2). Based on the fitted curve, we estimate the average of the reading, m, is 108.0 mmHg, and the standard deviation, σ, is 10.2 mmHg. From the normal distribution curve, we know that 68.27% of the readings will fall in the range of 108.0 ± 10.2 mmHg (97.8–118.2 mmHg), 95.45% in the range of 108.0 ± 2 × 10.2 mmHg (87.6–128.4 mmHg), and 99.73% in the range of 108.0 ± 3 × 10.2 mmHg (77.4–138.6 mmHg).

From these values, we are pretty confident that the person is not suffering from high blood pressure because there is only 0.14% chance for it to be so.

The average of 108.0 mmHg provides us with the best guess of the true blood pressure, while we are 99.73% confident that the true blood pressure is between 77.4 and 138.6 mmHg, which represents the repeatability or the range of the uncertainty. Unfortunately, we can never know what the true blood pressure is; we can only give our best guess and then decide based on how confident we are with our guess.

When we have additional information about a measurement, the implied uncertainty of using the significant figures and number of decimal places is no longer adequate. We should represent the measurement in the format of plus and minus, such as 108.0 ± 30.6 mmHg. If not specifically stated, we will assume that this is defined with 99.73% confidence. In many cases, a 95.45% confidence is sufficient; therefore, the measurement should be expressed as 108.0 ± 20.4 mmHg with 95.45% or simply 95% confidence. The confidence should be explicitly stated if it is not 99.73%.

1.2.3 Computation with uncertainties

We will use an excellent example and discussion from Doeblin (1990)[7] to demonstrate how to handle uncertainties in computation. Let a variable $y = f(x_1, x_2, x_3, \ldots, x_n)$, where x_i's denote the measured variables with the uncertainties $\Delta x_1, \Delta x_2, \Delta x_3, \ldots, \Delta x_n$. What will be the uncertainty Δy?

Using the Taylor series, we expand function f as,

$$
\begin{aligned}
y = y_0 + \Delta y &= f\left(x_1 + \Delta x_1,\, x_2 + \Delta x_2,\, x_3 + \Delta x_3, \ldots,\, x_n + \Delta x_n\right) \\
&= f\left(x_{10}, x_{20}, x_{30}, \ldots, x_{n0}\right) + \Delta x_1 \frac{\partial f}{\partial x_1} + \Delta x_2 \frac{\partial f}{\partial x_2} + \Delta x_3 \frac{\partial f}{\partial x_3} \\
&\quad + \ldots + \Delta x_n \frac{\partial f}{\partial x_n} + \text{H.O.T.}
\end{aligned}
\tag{1.1}
$$

where all the partial derivatives are to be evaluated at the known values of x_{i0}'s and the values of Δx_i's could be positive or negative. The variable y_0 is the nominal value of f with respect to x_{i0}'s. We can ignore the Higher Order Terms (H.O.T.) if the values of Δx_i's are small.

To determine Δy, we usually consider the worst-case and the best-case scenarios. The real case most likely falls somewhere in between.

For the worst-case scenario, we define the range of Δx_i as $R\Delta x_i$; therefore,

$-\frac{1}{2}R\Delta x_i < \Delta x_i < \frac{1}{2}R\Delta x_i$ and the range of Δy becomes,

$$
R\Delta y = \left(\left| R\Delta x_1 \frac{\partial f}{\partial x_1} \right| + \left| R\Delta x_2 \frac{\partial f}{\partial x_2} \right| + \left| R\Delta x_3 \frac{\partial f}{\partial x_3} \right| + \ldots + \left| R\Delta x_n \frac{\partial f}{\partial x_n} \right| \right)
\tag{1.2}
$$

We should express y as

$$
y = y_0 \pm \frac{1}{2} R\Delta y
\tag{1.3}
$$

---

[7]Doebelin, E.O. (1990). *Measurement Systems Application and Design.* McGraw-Hill Publishing Company, New York, NY. pp. 58–67., ISBN 0-07-017338-9.

If we consider Δx_i's as random variables with normal distributions, then we can compute the standard deviation of Δy as

$$\sigma_{\Delta y} = \sqrt{\left(\sigma_{\Delta x_1}\frac{\partial f}{\partial x_1}\right)^2 + \left(\sigma_{\Delta x_2}\frac{\partial f}{\partial x_2}\right)^2 + \left(\sigma_{\Delta x_3}\frac{\partial f}{\partial x_3}\right)^2 + \ldots + \left(\sigma_{\Delta x_n}\frac{\partial f}{\partial x_n}\right)^2}$$

(1.4)

From Equation (1.4), if we present the uncertainty of y with 99.73% confidence, we have

$$y = y_o + \Delta y = y_o \pm 3\sigma_{\Delta y}$$

(1.5)

Equation (1.5) represents the best-case scenario.

What if the uncertainties are not in normal distribution? In that case, there is no easy analytical way to determine Δy. However, with today's computing power, we can use the Monte Carlo simulation to determine the probability distribution of Δy and plot a histogram similar to that in Fig. 1.2.

Let us consider an example from Doeblin (1990) regarding the measurement obtained from a dynamometer. The output power of a dynamometer can be written as

$$P = \frac{2\pi RF(\frac{L}{12})}{550\,t} = \frac{2\pi}{550 \times 12}\frac{RFL}{t}$$

(1.6)

where P is the power in hp, R is the revolution of the shaft per second, F is the measured force in lbf at the end of the torque arm, L is the length of the torque arm in inches, and t is time in second, s.

For a specific run, if the data are

$$F = 10.12 \pm 0.040\,\text{lbf}$$

(1.7)

$$R = 1202 \pm 1.0\,\text{rev/s}$$

(1.8)

$$L = 15.63 \pm 0.050\,\text{in}$$

(1.9)

$$t = 60.0 \pm 0.50\,\text{s}$$

(1.10)

the uncertainties of these measurements are determined from the sensors' calibration. Note that all the uncertainty terms are expressed with one extra decimal place than the nominal value.

The nominal value of P is computed based on Equation (1.1). We then have

$$P_0 = \frac{2\pi}{550 \times 12} \frac{1202.0 \times 10.12 \times 15.63}{60.0} = 3.01668 \cong 3.02 \qquad (1.11)$$

We round the above value to two decimal places.

Now, we compute the partial derivatives in Equation (1.1) to three significant figures as

$$\frac{\partial f}{\partial x_1}|_0 = \frac{\partial P}{\partial F}|_0 = \frac{2\pi}{550 \times 12} \frac{LR}{t}|_0 = \frac{2\pi}{550 \times 12} \frac{15.63 \times 1202}{60.0} = 0.298\,\text{hp/lbf}$$
$$(1.12)$$

$$\frac{\partial f}{\partial x_2}|_0 = \frac{\partial P}{\partial R}|_0 = \frac{2\pi}{550 \times 12} \frac{FL}{t}|_0 = 0.00251\,\text{hp/r} \qquad (1.13)$$

$$\frac{\partial f}{\partial x_3}|_0 = \frac{\partial P}{\partial L}|_0 = \frac{2\pi}{550 \times 12} \frac{FR}{t}|_0 = 0.193\,\text{hp/in} \qquad (1.14)$$

$$\frac{\partial f}{\partial x_4}|_0 = \frac{\partial P}{\partial t}|_0 = \frac{-2\pi}{550 \times 12} \frac{FLR}{t^2}|_0 = -0.0500\,\text{hp/lbf} \qquad (1.15)$$

For the worst-case scenario, the ranges of uncertainties are based on Equations (1.7–1.10)

$$R\Delta F = 0.080\,\text{lbf} \qquad (1.16)$$

$$R\Delta R = 2.0\,\text{rev/s} \qquad (1.17)$$

$$R\Delta L = 0.100\,\text{in} \qquad (1.18)$$

$$R\Delta t = 1.00\,\text{s} \qquad (1.19)$$

From Equation (1.2), we have

$$R\Delta y = \left(\left| R\Delta F \frac{\partial P}{\partial F} \right| + \left| R\Delta R \frac{\partial P}{\partial R} \right| + \left| R\Delta L \frac{\partial P}{\partial L} \right| + \left| R\Delta t \frac{\partial P}{\partial t} \right| \right) = 0.098\,\text{hp}$$
$$(1.20)$$

From Equation (1.3), we have

$$P = P_0 \pm \frac{1}{2} R\Delta y = 3.02 \pm 0.049\,\text{(hp)} \cong 3.02 \pm 0.05\text{(hp)} \qquad (1.21)$$

For the best-case scenario, we first assume that the expressions of Equations (1.7–1.10) are based on 99.73% confidence level of normal distributions. Therefore, we have

$$\sigma_{\Delta F} = \frac{1}{3}(0.040)\text{lbf} \tag{1.22}$$

$$\sigma_{\Delta R} = \frac{1}{3}(1.0)\text{rev/s} \tag{1.23}$$

$$\sigma_{\Delta L} = \frac{1}{3}(0.050)\text{in} \tag{1.24}$$

$$\sigma_{\Delta t} = \frac{1}{3}(0.50)\text{s} \tag{1.25}$$

From Equation (1.4), the standard deviation of ΔP is

$$\sigma_{\Delta P} = \sqrt{\left(\sigma_{\Delta F}\frac{\partial P}{\partial F}\right)^2 + \left(\sigma_{\Delta R}\frac{\partial P}{\partial R}\right)^2 + \left(\sigma_{\Delta L}\frac{\partial P}{\partial L}\right)^2 + \left(\sigma_{\Delta t}\frac{\partial P}{\partial t}\right)^2}$$
$$= 0.00977\,\text{hp} \tag{1.26}$$

Finally,

$$P = P_0 \pm 3\sigma_{\Delta P} = 3.017 \pm 3\,(0.00977) \cong 3.017 \pm 0.029\,(\text{hp}) \tag{1.27}$$

In other words, we are confident (99.73% sure) that the actual value of P will lie between 2.988 and 3.046 hp. With a lower confidence of 95.45%, the true value of P will lie between 2.997 and 3.036 hp. There is only a 4.28% chance that the true value lies between 3.036 and 3.046 hp or between 2.988 and 2.997 hp. We use three decimal places for the best-case scenario.

1.2.4 Precision consideration in differential quantities

As shown in Equation (1.1), the higher order terms can be neglected if the values of Δx_i's are small. This is the case when we discuss the differential terms. For example, the Taylor expansion of $\sin\theta$ is

$$\sin\theta = \theta - \frac{\theta^3}{3!} + \frac{\theta^5}{5!} - \cdots \tag{1.28}$$

For a differential angle $d\theta$, it becomes

$$\sin d\theta = d\theta - \frac{d\theta^3}{3!} + \frac{d\theta^5}{5!} - \cdots \cong d\theta \tag{1.29}$$

when the H.O.T are neglected. What is the error if we neglect the high order terms? In this case, the error ratio, ε, defined as the error to the true

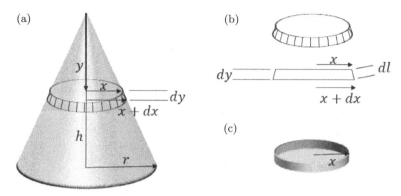

Figure 1.3:   (a) Volume calculation of a straight cone; (b) An infinitesimal conical layer; (c) An infinitesimal cylindrical layer.

value, is

$$\varepsilon = \frac{|sind\theta - d\theta|}{sind\theta} = \frac{\frac{d\theta^3}{3!} - \frac{d\theta^5}{5!} + \frac{d\theta^7}{7!} - \cdots}{sind\theta} < \frac{\frac{d\theta^3}{3!}}{sind\theta} \cong \frac{\frac{d\theta^3}{3!}}{d\theta} = \frac{d\theta^2}{3!} \qquad (1.30)$$

which is infinitesimal. Similarly, we found that $cosd\theta \cong 1$ and $tand\theta \cong d\theta$.

In Statics, there will be cases where integration will need to be done over different geometric shapes to determine the surface area, volume, and moments of inertia. Often, we have to decide on the inclusion or exclusion of differential terms. We will present a few cases for illustration.

As shown in Fig. 1.3, we want to determine the volume of a straight cone with a height h and a base radius r. We can slice the cone into infinitesimally thin disks and add up the volume of all the disks. The top surface of the thin disk in Fig. 1.3(b) has a radius x and the bottom surface has a radius $x + dx$. Let us say that we do not know how to determine the differential volume, dV, of the thin disk of (b), but we know it will be between those of two cylinders with a radius x and $x + dx$, respectively, as shown in (c). Therefore, we have

$$\pi x^2 dy < dV < \pi(x + dx)^2 dy \qquad (1.31)$$

If we decide to set $dV = \pi x^2 dy$, the error ratio, ε, will be

$$\varepsilon < \frac{\pi(x + dx)^2 dy - \pi x^2 dy}{\pi x^2 dy} = \frac{2dx}{x^2} + \frac{dx^2}{x^2} \qquad (1.32)$$

which is still negligible.

Now, integrating dV over y, we have the well-known result,

$$\int_0^V dV = \int_0^h \pi x^2 dy = \int_0^h \pi \frac{r^2}{h^2} y^2 dy = \frac{\pi}{3} r^2 h \qquad (1.33)$$

For the same straight cone, if we would like to determine the exterior cone surface, can we use the exterior surface of the thin cylindrical disk and carry out similar integration?

If we do so, then the exterior cone surface becomes

$$\hat{S} = \int_0^{\hat{S}} d\hat{S} = \int_0^h 2\pi x dy = \int_0^h 2\pi \frac{r}{h} y \, dy = \pi r h \qquad (1.34)$$

which is incorrect. So what went wrong?

The actual exterior surface of the thin disk in Fig. 1.3(b) should be

$$dS = 2\pi x dl = 2\pi x \sqrt{dx^2 + dy^2} = 2\pi \frac{r}{h} \sqrt{\frac{r^2 + h^2}{h^2}} y \, dy \qquad (1.35)$$

The error ratio between dS and $d\hat{S}$ is

$$\varepsilon = \frac{dS - d\hat{S}}{dS} = \frac{\sqrt{r^2 + h^2} - h}{\sqrt{r^2 + h^2}} = \frac{l - h}{l} \qquad (1.36)$$

where $l = \sqrt{r^2 + h^2}$, which is the length along the cone exterior surface. As shown in Equation (1.36), the error ratio is finite, not negligible. As a result, the error will accumulate, leading to mistakes. The correct answer is

$$S = \int_0^S dS = \int_0^h 2\pi x dl = \int_0^h 2\pi \frac{r}{h} \sqrt{\frac{r^2 + h^2}{h^2}} y \, dy = \pi r l \qquad (1.37)$$

Between $\hat{S}$ and S, we have the same error ratio as in Equation (1.36). A similar consideration should be applied when calculating the exterior surface of a paraboloid disk, as discussed in Problem 11.19 of Chapter 11.

1.3 Problem-Solving Protocol in Statics

To solve a problem in Statics, we should follow a set of systematic steps as follows:

(a) Problem statements;
(b) Force analysis;
(c) Governing equations;
(d) Solving equations;

(e) Answer verifications;

(f) Extensions.

In the problem statements, one should make proper sketches and define proper reference coordinates. List all the given conditions and the unknowns to be found.

For force analysis, we will present a systematic method in Chapters 5 and 6 so that the force analysis can be conducted correctly every time.

As regards governing equations, we typically construct force equilibrium equations based on the force analysis (Chapters 7 and 8). Additional governing equations can be found in Chapters 9 and 10.

As regards solving equations, after we have established relevant governing equations, we should always count the number of the unknowns associated with the governing equations. If the number of unknowns is greater than the number of the governing equations, we are not able to solve for all the unknowns. When this occurs, we should carry out the following steps to overcome this problem. First, we should check if there is an equation involving only one unknown. If so, this particular unknown can be solved. As a result, the number of unknowns is reduced by one, but the number of equations is also reduced by one. We should also check if there is a subset of the governing equations involving the same number of unknowns. If so, we can solve for those unknowns. Once we exhaust all the equations that can be solved, we still need to find additional equations.

To find additional equations, we should examine the unknowns and consider if they have specific relationships among them so that we can establish additional equations.

If we cannot find additional equations, then we have to make assumptions to define the values of some unknowns. For example, in the case of a smooth surface the friction can be assumed to be zero.

Finally, as engineers, we should know that we can conduct measurements using sensors to determine the values of some unknowns. Once we reduce the number of unknowns to be the same as that of the governing equations, we can proceed to solve for all the unknowns.

In solving equations, we can use computing aids for help. Today, there are equation solvers useful for solving complicated equations, numerically or symbolically. We provide a few programming examples in this book.

After we have solved the equations and obtained the values of unknowns, we should conduct answer verifications. Do they look reasonable? We could have made some mistakes along the way. Do they violate the assumptions? Do they violate basic laws?

Finally, we should think about the implication of the results for extensions.

1.4 Concluding Remarks

Practice makes perfect. It applies to both honing the problem-solving skill and deepening the understanding of the concept. In Chapter 11, we provided detailed solutions to 31 difficult problems based on Appendix A of the classic Statics textbook by Meriam.[8] These problems, along with the examples presented in Chapters 1–10 are useful to learn how to think properly to solve difficult Statics problems through proper understanding of difficult concepts in Statics.

---

[8]Meriam, J.L. (1975). *Statics*, 2nd Edition. John Wiley & Sons, Inc., Hoboken, NJ.

Chapter 2

What is a Force?

The entire subjects of Statics and Dynamics basically deal with the effects or consequences of forces. But first, what is a force? This is not an easy question to answer. In general, we do not observe a force directly, only its effects. The concept of a force was first more scientifically defined based on its effect on the motion of an object, which is, of course, *Newton's Second Law*,

$$\sum \vec{F} = m\vec{a} \qquad (2.1)$$

The left-hand side of this equation, $\sum \vec{F}$, represents the summation of many forces acting on an object, while the right-hand side, $m\vec{a}$, is related to the effects of the force on the object, where m is the mass of the object and $\vec{a}$ is related to the motion of the object, or more precisely, the acceleration of the object. This famous, apparently simple, equation actually contains three drastically different subjects of physics, and each of them could warrant a lifelong study to uncover their fundamental roots. However, for the purpose of basic engineering, we will only address them from the point of applying them for solving engineering problems.

Let us also agree for the time being that *Newton's Second Law* is correct. Basically, we have three unknowns involved in this equation. We must know two of them in order to determine the third unknown based on Equation (2.1). If we are interested in determining the motion of the object, we must independently determine the quantities of the left side, the forces, and that of the mass, in order to determine the motion.

Therefore, the question now becomes, "How do we determine a force independently?" Before trying to answer this question, let us first examine different types of forces.

2.1 Contact and Non-Contact Transmitted Forces

Let us agree first that we will address some specific "influences" which cause some specific effects of objects as forces. By observing these effects, we can define different types of forces. In Physics, there are forces involved within the molecular and atomic structures, which exert influences on how the atoms bond together, how the electrons orbit around nucleus, etc. These forces are beyond the scope of basic engineering subjects regarding Statics and Dynamics. For basic engineering, it is useful to put forces into two different categories based on how a force is applied to an object. If a force must be exerted on an object by putting it in contact with or bonding with another object, the force is considered as a contact-transmitted force; otherwise, it is a non-contact transmitted force.

The non-contact transmitted forces are fundamental forces due to physical fields, such as gravitational fields, electric fields, and magnetic fields. Among them, the gravitational field is most important to mechanical systems and will be discussed in more detail in the next section. When a satellite is orbiting in the space around the Earth, it is apparently not in direct contact with any other object, but a gravitational force can still act on it by the Earth; therefore, gravitational forces are non-contact transmitted forces.

A contact transmitted force can only be applied when an object is in contact with another object. The word "contact" is defined very broadly here, and we will define it more precisely later in this section.

Contact transmitted forces are more obvious to us. If you bang your head against a wall, you feel a "force" at the point where your head is in contact with the wall. Because you can remove your head off the wall readily, the "contact" condition between the head and the wall is denoted as "simple *contact*", or just "*contact*". On the other hand, if you pull your right hand index finger lightly with your left hand, you can feel the force exerted by your left hand on the index finger through the simple contact. Because your index finger is not pulled off your right palm, you also feel another force which keeps your right index finger from coming off. Technically, the finger and the palm are not just in contact, but joined together via muscle and tendon. They cannot be readily separated. In other words, they are bonded together, not just in a simple contact. We will call this more complicated contact condition as *constraint*. It becomes clear that contact transmitted forces can only be transmitted through simple contacts or constraints.

One important contact transmitted force is the frictional force. If you rub your face with your hand, you feel this force because your hand is in contact with the face. When you are driving, the air particles are in contact with the car, exerting a contact force, called the wind drag. Because the air is not visible, we often forget that we are in contact with the air all the time. Frictional forces are very important, and many engineering solutions are devised to overcome them. However, in most standardized problems of textbooks in Statics and Dynamics, air resistance and friction are often neglected without proper justification. One could be mistaken that we live in a giant vacuum and a frictionless world from those assumptions.

2.2 Mathematical Representation of Forces

When math joins hands with physics, it opens up new ways to understand physics. Therefore, we should define useful math tools to describe forces.

Judging the effect of a force, it is observed that a force is directional, with a magnitude or size, and acting on a specific point of an object, as shown in Fig. 2.1. Therefore, mathematically, a force can be represented by a vector. Typically, we use $\vec{F}$ to represent a force. Here we use a vector sign over a letter to represent a vector, instead of using a bold-face letter. This is done because when we solve problems by hand, we write it as $\vec{F}$, not $\boldsymbol{F}$. It is very hard to distinguish F from $\boldsymbol{F}$ in handwriting. Therefore, in this book,

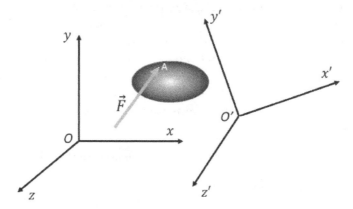

Figure 2.1: A force vector $\vec{F}$ defined by different reference frames.

we opted to use $\vec{F}$ instead of $\boldsymbol{F}$. Mathematically, the direction of vector $\vec{F}$ can be determined based on a reference frame, such as the $O - x - y - z$ or the $O' - x' - y' - z\prime$ coordinate in Fig. 2.1.

It should be noted that one can choose any reference frame, $O - x - y - z$ or $O' - x' - y' - z\prime$, to describe a force, but the size and the direction of the force do not change because of the choice of the reference frame. As shown in Fig. 2.1,

$$\vec{F} = a\vec{i} + b\vec{j} + c\vec{k} = a'\vec{i'} + b'\vec{j'} + c'\vec{k'} \qquad (2.2)$$

where $\vec{i}, \vec{j}, \vec{k}$ and $\vec{i'}, \vec{j'}, \vec{k'}$ are unit vectors of the reference frames, $O - x - y - z$ and $O' - x' - y' - z'$, respectively. The directions and sizes of these unit vectors may or may not be the same. Typically, we only have to choose one convenient reference frame for force analysis. For the moment, there is no concern regarding the properties of the reference frame, such as if the origin of the reference frame is moving or not or if the unit vectors could be changing in their directions. In fact, the force observation should be independent of the reference frame, and we only use the unit vectors for representing directions.

For common conventions, when designating a Cartesian reference frame, we need to follow the right hand rule. If we place four fingers of the right hand to point to the positive direction of the x-axis, and then fold the fingers naturally toward the palm to align with the positive direction of the y-axis, the thumb then points to the positive direction of the z-axis. We cannot arbitrarily assign the axes of a reference frame. For example, y and z cannot be switched in Fig. 2.1.

Once we have selected a reference frame to define a force as a vector, we need to specify the component of the force along each axis. The best way to express these components is by the angles between the force vector and the axes. There are several ways of doing this. For a 2D case, as shown in Fig. 2.2(a), we can define an angle, θ, between $\vec{F}$ and $\vec{i}$, which is the unit direction vector of the x-axis. The magnitude of $\vec{F}$ is $|\vec{F}|$, and graphically, it is the length of the vector arrow. The vector expression of $\vec{F}$ becomes

$$\vec{F} = |\vec{F}|(\cos\theta\vec{i} + \sin\theta\vec{j}) = |\vec{F}|(\cos\theta\vec{i} + \cos(90^\circ - \theta)\vec{j}) \qquad (2.3)$$

where $(90^\circ - \theta)$ is the angle between $\vec{F}$ and $\vec{j}$.

The second expression of Equation (2.3) provides a convenient form when we extend it to a 3D case, as shown in Fig. 2.2(b). The direction vector is simply the cosine of the angle of $\vec{F}$ with respect to each axis;

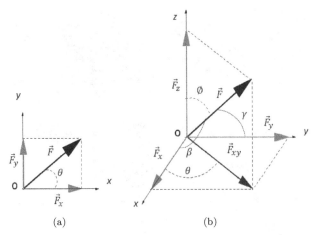

Figure 2.2: A force can be broken down into their components: (a) a 2D case and (b) a 3D case.

therefore,

$$\vec{F} = |\vec{F}|(\cos\beta\vec{i} + \cos\gamma\vec{j} + \cos\phi\vec{k}) \qquad (2.4)$$

where β is the angle between $\vec{F}$ and $\vec{i}$, γ is the angle between $\vec{F}$ and $\vec{j}$, and ϕ is the angle between $\vec{F}$ and $\vec{k}$.

The vector, $\cos\beta\vec{i} + \cos\gamma\vec{j} + \cos\phi\vec{k}$, is denoted as direction cosine.

Although Equation (2.4) is simple, the angles, β, γ, and ϕ are not easily measured in practice. These angles must be measured on the basis of the individual plane formed by $\vec{F}$ and the corresponding axis (see Fig. 2.2(b)). These planes are different for different $\vec{F}$. Practically, it is difficult to mount a sensor that can be used to measure an angle on different planes.

The most practical sensor for measuring angles is a shaft encoder, as shown in Fig. 2.3. If this encoder is used to measure the rotation around the x-axis, its measurement plate will be perpendicular to the x-axis.

In practice, we can install a shaft encoder aligned with the z-axis, which will measure the angle θ shown in Fig. 2.2(a), on the x–y plane. Angle θ represents the angle between $\vec{F}_{xy}$ and the x axis. $\vec{F}_{xy}$ is the projection of $\vec{F}$ on the x–y plane. We can install a second shaft encoder, perpendicular to $\vec{F}_{xy}$, to measure the angle between $\vec{F}_{xy}$ and $\vec{F}$, on the plane defined by $\vec{F}_{xy}$ and $\vec{F}$. Technically, this can be done without much difficulty. This angle is $90° - \phi$, while ϕ is the angle between $\vec{F}$ and the z-axis, as defined before.

Now we have a different expression for $\vec{F}$ based on θ and ϕ,

$$\vec{F} = |\vec{F}|(\sin\phi\cos\theta\vec{i} + \sin\phi\sin\theta\vec{j} + \cos\phi\vec{k}) \qquad (2.5)$$

Figure 2.3:   A shaft encoder, which is a disk with radial strips for angle marking and a pick-up sensor. An enhanced image based on the image from https://arduino.stackexchange.com/questions/11962/how-to-read-data-from-a-rotary-encoder-with-atmega328.

Equation (2.5) is more practical, requiring only two sensors, but its expression is more complex, while Equation (2.4) is just the opposite. Typically, Equation (2.5) is preferred.

Finally, how do we determine $|\vec{F}|$? If the force is expressed as Equation (2.2), then

$$|\vec{F}| = \sqrt{a^2 + b^2 + c^2} = \sqrt{a'^2 + b'^2 + c'^2} \qquad (2.6)$$

The expression in Equations (2.4) or (2.5) is very convenient because $|\vec{F}|$ represents the size, while $(\sin\phi\cos\theta\vec{i} + \sin\phi\sin\theta\vec{j} + \cos\phi\vec{k})$ is the direction unit vector for $\vec{F}$ because its magnitude is 1 and its direction is the same as $\vec{F}$.

We can further generalize the force vector as

$$\vec{F} = |\vec{F}|\vec{e}_F \qquad (2.7)$$

where $\vec{e}_F = (\sin\phi\,\cos\theta\vec{i} + \sin\phi\sin\theta\vec{j} + \cos\phi\vec{k}) = (\cos\beta\vec{i} + \cos\gamma\vec{j} + \cos\phi\vec{k})$, representing the directional unit vector of $\vec{F}$. Most importantly, if we know $\vec{F}$, $\vec{e}_F$ can be easily calculated as

$$\vec{e}_F = \frac{\vec{F}}{|\vec{F}|} \qquad (2.8)$$

2.3 Combining the Effects of Forces

The vector mathematical operations include vector addition, subtraction, both dot and cross multiplications, etc. Once forces are represented by vectors, many basic vector operations can be applied to forces to determine $\sum \vec{F}$. However, there are a few additional rules to be considered for the physics of forces.

The physics of forces requires that we consider where the force is applied, denoted as the point of application, and the line of action, along which the force is heading, passing through the point of application. The property of the object on which the force is applied should be considered as well.

In general, in order to combine two forces, both forces must be acting on the same point of application. As shown in Fig. 2.4, two forces, $\vec{F}_1$ and $\vec{F}_2$, are acting on the same point, A. We have the choice of representing the force acting on point A as pointing at point A (Fig. 2.4(a)) or starting from point A (Fig. 2.4(b)). We simply need to choose one way and be consistent with it. We can then combine these two forces to form a new force, $\vec{F}$, using the vector addition. Graphically, the addition is carried out as shown in Fig. 2.4. Mathematically, it is simpler. We simply add up each component.

$$\vec{F} = \vec{F}_1 + \vec{F}_2 = (a_1\vec{i} + b_1\vec{j} + c_1\vec{k}) + (a_2\vec{i} + b_2\vec{j} + c_2\vec{k})$$
$$= (a_1 + a_2)\vec{i} + (b_1 + b_2)\vec{j} + (c_1 + c_2)\vec{k} \tag{2.9}$$

The reason that two forces not applied at the same point could not be combined can be explained with practical experience. As shown in Fig. 2.5, if we push our left and right cheeks in with fingers, the force applied by right index finger cannot be combined with the one by the left finger. This is because each force creates its own local depression, and so the two cannot be combined. For many engineering applications, these local effects may not

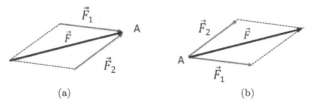

(a)   (b)

Figure 2.4: Two forces acting on the same point can be combined as vectors into one single force: (a) with forces pointing to point A and (b) with forces originating from point A. Both ways are correct.

Figure 2.5:   Forces applied to the human face cannot be combined as vectors.

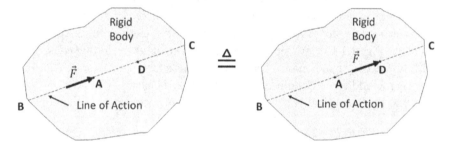

Figure 2.6:   A force can move along the line of action on a rigid body without changing its effect on the rigid body.

be significant if the object is very rigid. In an extreme case, if we assume that the object is so rigid that there is no local effect whatsoever, then we can call this kind of object as a rigid body. Therefore, a rigid body never deforms, no matter how many or how big the forces applied to it.

This assumption of a rigid body is valid in many engineering applications when the deformation is negligible due to the strength of the material used for the object. For example, a steel nail is considered as rigid when you hammer it into a piece of wood, while the wood is not a rigid body.

Once an object can be assumed as a rigid body, then a force can be moved along its line of action without changing its effect on the rigid body, which is explained in Fig. 2.6.

The line of action of a force starts and ends on the rigid body. For Fig. 2.6, the line of action of $\vec{F}$ starts at point B and ends at point C. The same force can be applied at point A or D, both on the same line of action,

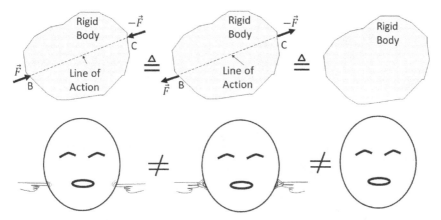

Figure 2.7: A pair of forces with equal size and opposite directions can be cancelled on the rigid body, but not on a non-rigid body.

and the effects on the rigid body are the same. With this reasoning, we have a bit of a silly situation as shown in Fig. 2.7.

As can be seen from Fig. 2.7, a pair of compressive forces has the same effect on a rigid body as a pair of tensile forces, as well as no forces, because these pairs of forces can be cancelled if they are the same size after moving them along the line of action. Apparently, this is not the case on a non-rigid human face. The cheeks could be pulled out, snapped in, or no changes can occur with different force conditions. This silly example reminds us of the limit of the rigid body assumption.

The assumption of rigid body is important. If it is a reasonable assumption, we can now combine forces even though they are not applied at the same point of application. Any two forces, not in parallel, now can be moved along their respective lines of action to the point where these two lines of action intersect, if they do have an intersection point on the rigid body. For a pair of forces, with equal size, but in opposite directions, and not on the same line of action, we apparently cannot combine them, as shown in Fig. 2.8. The combined action of such a force pair causes a rotation of the rigid body. This force pair is denoted as a couple or a force couple. This rotation action is described by the product of the size of the forces and the distance between the two lines of action. As in the case of Fig. 2.8, this force couple, which forms a moment, denoted as $\vec{\tau}$, is in clockwise rotation. Based on the right hand rule, the clockwise rotation is pointing to the direction of $-\vec{k}$.

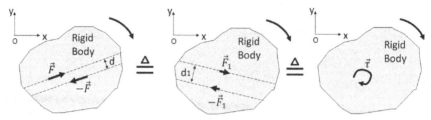

Figure 2.8: A pair of forces with equal size and opposite directions can be cancelled on a rigid body if they are on the two parallel lines of action. The combined action is a rotating effect.

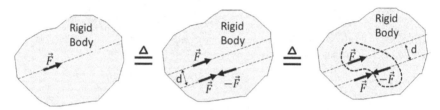

Figure 2.9: A force can be moved to a new parallel line of action, but it would result in an accompanying torque.

It turns out that different force pairs can form the same moment as $\vec{\tau}$, as shown in Fig. 2.8 and Equation (2.10).

$$\vec{\tau} = (|\vec{F}| \cdot d)(-\vec{k}) = (|\vec{F_1}| \cdot d_1)(-\vec{k}) \tag{2.10}$$

This moment $\vec{\tau}$ does not have a point of application on a rigid body. We can apply a moment anywhere on the rigid body, and its rotating action is the same. In other words, $\vec{\tau}$ is a free vector and it can be anywhere on the rigid body as long as the direction and size are kept the same. Of course, this free vector property does not exist for non-rigid bodies. For example, if you twist your shirt at different spots, you would create different wrinkles.

Finally, we realize that we can also move a force out of its line of action, but there will be a consequence of a force couple. This result is explained in Fig. 2.9. To move a force to a new line of action, we first create a null pair of forces: two forces on the same point with equal size but in opposite directions. The magnitude of the null force pair is the same as that of the original force. This null pair essentially is zero; therefore, adding this null pair does not change the loading condition of the rigid body. If we let one force of the null pair form a force couple with the original force vector,

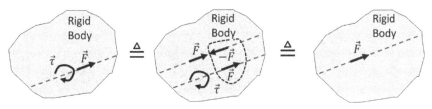

Figure 2.10: All forces can be combined into one force and one moment, and further simplified to just one force along a specific line of action on a 2D rigid body.

the remaining force of the null pair becomes the original force but on a new line of action. The force couple (the moment) is free to be anywhere on the rigid body. The new force line of action and the moment have the same effect on the rigid body as the original force on the original line of action.

Based on the results of Fig. 2.9, we can move any force acting on a rigid body so that all of them act at the same point. Once they are at the same point, they can be combined as vectors. All the resulting torques can also be added together as vectors. Finally, we conclude that on a 2D rigid body, all the forces can be combined to become a total force and a total moment on a specific line of action as shown in Fig. 2.10. We can then move this total force to a different line of action so that the resulting moment can cancel the total moment. The end result is that all forces on a 2D object can be combined into one total force acting on a specific line of action. A special case of this result is when an object is subjected to gravitational forces on every particle of the rigid body. All these forces can be combined into a total force acting on the mass center of the object. The mass center concept will be explained in later chapters.

On a 3D rigid body, the case is a bit more complicated because the resulting force and moment may not be perpendicular as in the case of a 2D rigid body. However, as shown in Fig. 2.11, we can always split the moment into two components: one aligned with the total force, $\vec{\tau}_W$, and the other perpendicular to the total force, $\vec{\tau}_M$. The perpendicular component can be nulled by moving the total force to a different line of action, similar to Fig. 2.10; however, the moment component aligned with total force, $\vec{\tau}_W$, cannot be nulled. The aligned moment, $\vec{\tau}_W$, is called a wrench because it is similar to using a screwdriver to tighten or loosen a screw.

Combining all the forces and moments acting on a rigid body to a total force and a wrench appears to simplify the force representation, but this has very little practical value. One main reason is that most objects are not

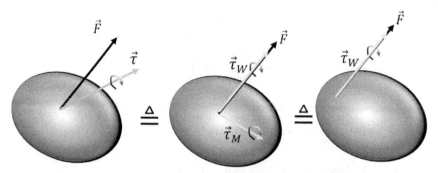

Figure 2.11: On a 3D rigid body, all forces can be combined into one force and one moment, and further simplified to just one force with a wrench along a specific line of action.

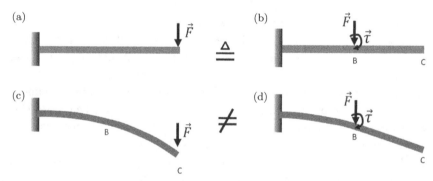

Figure 2.12: For rigid bodies as in (a) and (b), we can move a force accordingly. However, for a non-rigid body, even if the force conditions are equivalent, the resulting deformation of the non-rigid body is different. Note that the section of BC of beam (c) is curved, while the same section of beam (d) is straight.

rigid bodies and we cannot combine all the forces and moments by moving forces and moments to the same position and combining them.

Figure 2.12 illustrates why we cannot move forces and moments on a non-rigid body as in a rigid body. For a rigid beam, as in Figs. 2.12(a) and 2.12(b), we can move the force accordingly. However, for a non-rigid beam, as in Figs. 2.12(c) and 2.12(d), the beam will deform differently even if the force and moment conditions are the same. In particular, in Fig. 2.12(c), section BC of the beam is curved, while in Fig. 2.12(d), it is straight.

This actually brings up a fundamental question regarding if we can combine forces and moments in general. This question will be answered in Chapter 3 when we discuss the condition when a force equilibrium is reached. For the time being, it is worthwhile to pay attention to the rigid body assumption so that we do not have a wrong impression that we can move a force at will.

2.4 How to Calculate Moments

The method illustrated by Fig. 2.9 to calculate a moment is not very convenient. There are simpler ways if we use vector math. We shall start with 2D cases and proceed to 3D cases.

As shown in Fig. 2.13, the resulting moment can be calculated with the following equation,

$$\vec{\tau}_A = \vec{r}_{B/A} \times \vec{F} \tag{2.11}$$

The notations used here are important and should be defined clearly. $\vec{\tau}_A$ is defined as the resulting moment if we move $\vec{F}$ from its original position B to point A. Because we do not want to actually move the force to determine $\vec{\tau}_A$, we can only state that $\vec{\tau}_A$ is the moment of $\vec{F}$ with respect to point A. $\vec{r}_{B/A}$ is a position vector of point B with respect to point A. Note that B/A implies B with respect to A. Graphically, we draw an arrow from point A to point B for $\vec{r}_{B/A}$. To determine a position vector, the easiest way is to use the position coordinates of the points. Therefore,

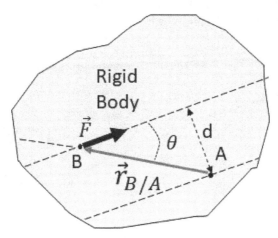

Figure 2.13: The vector analysis for calculating the moment of a force with respect to a point.

$$\vec{r}_{B/A} = \vec{r}_{B/O} - \vec{r}_{A/O} \tag{2.12}$$

For a 2D case, $\vec{\tau}_A$ of Fig. 2.13 has a direction, based on the right-hand rule, as pointing into the paper, while the size of $\vec{\tau}_A$ is

$$|\vec{\tau}_A| = |\vec{r}_{B/A}|\,|\vec{F}|\sin(\pi - \theta) = |\vec{r}_{B/A}|\,|\vec{F}|\sin(\theta) = d\,|\vec{F}| \tag{2.13}$$

The result of Equation (2.13) is the same for any force on the same line passing through point B. Therefore, for 2D cases, we can use the right hand rule to determine the direction of the resulting moment.

However, for 3D cases, the right hand rule is very hard to use. It would be much easier if we simply use the vector multiplication of Equation (2.11) to determine the moment.

With respect to Fig. 2.14, for example, if a force, $\vec{F} = 4\vec{i} - 2\vec{j} - 3\vec{k}$, acts at point B, the moment $\vec{\tau}_A$ due to $\vec{F}$ with respect to point A is

$$\vec{r}_{B/A} \times \vec{F} = \begin{vmatrix} \vec{i} & \vec{j} & \vec{k} \\ -2 & 4 & 5 \\ 4 & -2 & -3 \end{vmatrix} = -2\vec{i} + 14\vec{j} - 12\vec{k} \tag{2.14}$$

If $\vec{F}$ is not given, but the size is given with its direction expressed graphically, we can conveniently find two points along the force vector direction and determine a vector along this force direction. The unit force vector direction can then be determined similar to Equation (2.8). For example, if the force vector with a size 5 (we will discuss the unit in the next chapter) passes through point C with a coordinate, $(1, 3, 6)$, and point D,

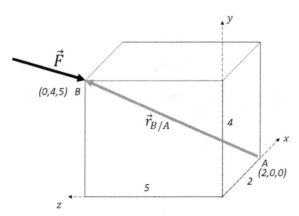

Figure 2.14: The vector analysis for calculating the moment of a force with respect to a point for a 3D case.

$(3, -2, 1)$, and is pointing from C to D, the unit direction vector of $\vec{F}$ is

$$\vec{e}_F = (\vec{r}_{D/C})/|\vec{r}_{D/C}| = \frac{(2\vec{i} - 5\vec{j} - 5\vec{k})}{\sqrt{2^2 + (-5)^2 + (-5)^2}} = \frac{(2\vec{i} - 5\vec{j} - 5\vec{k})}{\sqrt{54}} \qquad (2.15)$$

Therefore,

$$\vec{F} = |\vec{F}|\vec{e}_F = \frac{5(2\vec{i} - 5\vec{j} - 5\vec{k})}{\sqrt{54}} \qquad (2.16)$$

2.5 The Use of Dot Multiplication

The direction vector, $\vec{e}_F$, is very useful. We can use dot multiplication to determine the component of a force vector along any direction. For example, for a force vector, $\vec{F} = a\vec{i} + b\vec{j} + c\vec{k}$, the component of $\vec{F}$ in the $\vec{i}$ direction can be determined as follows:

$$a = \vec{F} \cdot \vec{i} = (a\vec{i} + b\vec{j} + c\vec{k}) \cdot \vec{i} \qquad (2.17)$$

Similarly, we can find out the components in $\vec{j}$ and $\vec{k}$ directions. Let us revisit Fig. 2.11 to show how we can split the total moment vector, $\vec{\tau}$, into two components: one aligned with the total force, denoted as wrench ($\vec{\tau}_W$) and the other perpendicular to the total force ($\vec{\tau}_M$).

We can easily determine the direction vector, $\vec{e}_F$, of $\vec{F}$ using Equation (2.8). Once $\vec{e}_F$ is determined, $\vec{\tau}_W$ can be determined easily as

$$\vec{\tau}_W = (\vec{\tau} \cdot \vec{e}_F)\vec{e}_F \qquad (2.18)$$

and

$$\vec{\tau}_M = \vec{\tau} - \vec{\tau}_W \qquad (2.19)$$

2.6 Concluding Remarks

Readers are encouraged to apply the concepts and the methods presented in this chapter to the examples or homework problems given in any textbook on Engineering Statics. In Chapter 11, there are a few problems for readers to practice, such as Problems 11.5, 11.7, and 11.11. The old saying of "practice makes perfect" is very true. Another important exercise is related to memory capacity. Although 19 equations are presented in this chapter, the reader should not try to memorize any of these equations. Equations with more complicated forms should be derived with proper graphics, instead of purely by memorization. For example, Equations (2.3–2.5) should be derived based on Fig. 2.2. Therefore, one should study this figure closely and be able to reproduce it from reasoning. Useful equations deserve some commitment by memory, such as Equations (2.7–2.8) and (2.11). Finally, one should practice the mental skill to solve the problem from scratch, without prior knowledge of solutions. Try not to remember how to solve a specific problem, but use the general strategy to solve problems in a systematic fashion, to be presented in later chapters.

Chapter 3

How Do We Quantify a Force?

As discussed in Chapter 2, we usually do not observe a force directly, only its effects. Therefore, to quantify or measure a force, we need to observe the effects of the force. Newton is basically the architect of the concept of force. His *second law*, $\sum \vec{F} = m\vec{a}$, describes how forces can be measured by measuring the changes of motion (acceleration) together with the mass of the object. A more general concept is that a force will change the momentum of an object. In Statics, the object is at equilibrium, and we only have $\sum \vec{F} = 0$. Apparently, there is no motion observation to provide us the force information. So, we have to find a different way.

3.1 Newton's Law of Gravitation

Other than measuring forces through *Newton's Second Law*, Newton also predicts the attractive force between two objects with masses based on his *Law of Gravitation*,

$$|\vec{F}| = G\frac{m_1 m_2}{r^2} \tag{3.1}$$

where r is the distance between two objects with masses m_1 and m_2, respectively, and G is the universal constant of gravitation. Equation (3.1) actually predicts two forces, as shown in Fig. 3.1, in opposite directions with the same magnitude on the same line. We will call it line of direction, not line of action because line of action is defined on a rigid body. The line of direction passes through the mass centers of these two objects. The concept of the mass center will be discussed later.

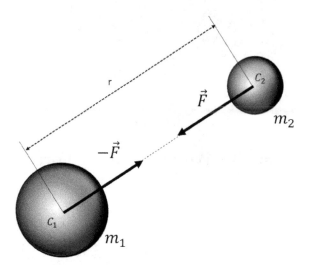

Figure 3.1:   Newton's Law of gravitation.

Note that these two forces, $-\vec{F}$ and $\vec{F}$, cannot be added as vectors and cancel each other, unlike the case of rigid bodies discussed in Chapter 2, because they do not act on the same point. We cannot move the forces along the line of direction (not line of action) so that they can meet at the same point to cancel each other.

With the *Law of Gravitation*, we no longer need to measure the acceleration of an object to determine the amount of the force acting on it. We can simply predict the attractive force between two bodies with the information of their masses and the distance between them.

The attractive force exerted by the Earth to pull an object of mass m is

$$F = G\frac{m\,M}{R^2} = m\left(G\frac{M}{R^2}\right) = mg \tag{3.2}$$

where M is the mass of the Earth and R is the distance of the object from the center of the Earth. If the object is near or on the surface of the Earth, R is essentially the same; as a result, the term $G\frac{M}{R^2}$ is a constant, denoted as g for the gravitational acceleration near the Earth's surface.

Equation (3.2) describes the size of the gravitational force between the Earth and an object. For the force applied to the object by the Earth, the direction of the force is pointing toward the center of the Earth as shown in Fig. 3.2. From our local view, the surface of the Earth is flat and we can assign an x–y reference frame to define the force direction. As a result, the direction of the force applied by the Earth to the object is

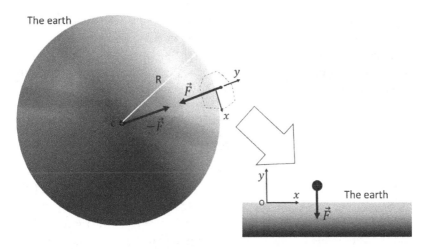

Figure 3.2: Gravitational force on the surface of Earth.

pointing in the $-\vec{j}$ direction based on this local reference frame, defined in Fig. 3.2. The form of Equation (3.2) resembles the syntax of *Newton's Second Law*. Therefore, g has the same meaning of acceleration and is denoted as gravitational acceleration. Because g is an acceleration, it has a direction; so, we should express it as $\vec{g}$ with its direction pointing toward the center of the Earth.

Now, we can predict how much force the Earth applies to any object near the surface of the Earth as

$$\vec{F} = m\,\vec{g} \tag{3.3}$$

The value of g is found to be, on average, $9.81\,\mathrm{m/s^2}$ or $32.2\,\mathrm{ft/s^2}$. These values are average values. If you are at the peak of a high mountain, the value of g will be less. There are other factors affecting the value of g due to the rotation of the Earth. According to Stevens and Lewis (2003),[1] the value of g is $9.832\,\mathrm{m/s^2}$ at the North Pole and $9.780\,\mathrm{m/s^2}$ at the equator, about 0.5% change, or $\pm 0.25\%$ from the average value of $9.81\,\mathrm{m/s^2}$.

The average value defined above is sufficiently accurate in most engineering calculations.

Note that the value of $g = G\frac{M}{R^2}$ rapidly decreases as the distance R becomes larger. The average radius of the Earth is 6,371 km. The distance between the Moon and the Earth is 384,400 km. The "g" value which the

---

[1]Stevens, B.L. and Lewis, F.L. (2003). *Aircraft Control and Simulation*, 2nd Edition. Hoboken, New Jersey: John Wiley & Sons, Inc. ISBN 0-471-37145-9.

Moon experiences from the Earth is 0.03% of the g we experience at the surface of the Earth. Therefore, if we are at the Moon's surface, we will experience very little pull from the Earth but a lot more pull from the Moon. The "g" value at the surface of the Moon is found to be 16.6% that of the Earth, based on the mass of the Moon being 1/81th that of the Earth and its radius being 1,738 km or 1/3.67th that of the Earth. Therefore,

$$g_{\text{moon}} = \frac{G\left(\frac{1}{81}M\right)}{\left(\frac{1}{3.665}R\right)^2} = 0.166\, g_{\text{earth}} = 1.63\,\text{m/s}^2 \tag{3.4}$$

When we watch films of astronauts in the International Space Station (ISS), they appear to be weightless. One would think that there is very little gravitational pull by the Earth. However, the average orbit radius of ISS is about 400 km above the surface of the Earth. The "g" value at ISS is

$$g_{\text{ISS}} = \left(G\frac{M}{(6371 + 400)^2}\right) = 0.885\, g_{\text{earth}} \tag{3.5}$$

In fact, the astronauts are still subjected to nearly 90% of the gravitational pull we experience on the surface of the Earth. How come they appear to be weightless or in a so-called a "microgravity" (nearly zero gravity) environment? Astronauts can lift a "heavy" object effortlessly in ISS. We cannot answer this question until we discuss the subjects of Dynamics.

The readers might have noticed that the word "weight" is not used in the above discussions. The concept of weight will be introduced later, and it is not exactly the same as the gravitational pull applied to an object by the Earth.

3.2 Units of Forces

Based on *Newton's Second Law*, we can define the unit of force. If we define a force based on International System of Units (SI Units), the mass is defined in kilogram (kg), the time in second (s), and the length in meter (m). The length unit is also called the metric system. The acceleration takes on the unit of m/s^2.

The unit for the force becomes kg m/s^2. However, this unit is too cumbersome to write. Therefore, the SI Units conveniently define

$$1\,N = (1\,\text{kg})\left(1\frac{\text{m}}{\text{s}^2}\right) = 1\left(\text{kg}\frac{\text{m}}{\text{s}^2}\right) \tag{3.6}$$

where N stands for the *newton* as a new unit for the force. Equation (3.6) is conveniently defined and can be described as: if we observe an object with 1 kg of mass undergoing $1\,\text{m/s}^2$ of acceleration, then the amount of the net force applied to the object is 1 N. The same object if placed on the surface of the Earth, based on Equation (3.3), will be subjected to a gravitational pull by the Earth that is

$$\vec{F}_w = m\,\vec{g} = (1\,\text{kg})(9.81\,\text{m/s}^2) = 9.81\,\text{kg}\,\text{m/s}^2 = 9.81\,\text{N} \qquad (3.7)$$

Equation (3.6) can be rewritten as

$$1 = \frac{\text{N}}{(\text{kg}\,\text{m/s}^2)} = \frac{(\text{kg}\,\text{m/s}^2)}{\text{N}} \qquad (3.8)$$

Equation (3.8) is denoted as a unity equation. It is very useful for unit conversion because we can multiply a unity equation to any equation without altering it. For example, if we want to change the unit of $\vec{F}_w$ of Equation (3.7) from $\text{kg}\,\text{m/s}^2$ to N, we can proceed as follows:

$$\vec{F}_w = 9.81\,\text{kg}\,\text{m/s}^2$$

$$= 9.81\,\text{kg}\,\text{m/s}^2 \times 1 = 9.81\,\text{kg}\,\text{m/s}^2 \times \frac{\text{N}}{(\text{kg}\,\text{m/s}^2)} = 9.81\,\text{N} \qquad (3.9)$$

In fact, we can define many different forms of Equation (3.8) to convert units.

Other commonly used units, in particular in the US, are based on the foot (ft), the pound mass (lb_m), and the second (s). Quite often, these units are referred to as English units in the US, but they are more customarily used in the US than in the UK, in particular for length. If a person walks into a tool store in the US to buy hand tools, he or she might be asked: "Do you want standard or metric?" The "standard" here means the length system in inches and feet. The standing joke is that this "standard" is not the standard used in the world. Moreover, it should not to be referred to as the English unit either because, in the UK, meters are used. Therefore, the metric system should be referred to as the standard, while the *"standard"* should be referred to as "American". The question that should be asked is "Do you want standard or American?"

Another confusion with this unit system is related to the word "pound". In this book, we should be very precise in indicating which "pound" we are referring to because the same word is used to indicate either mass or force in most standard textbooks in Statics and Dynamics. This causes a lot of confusion, and it can result in major mistakes in numerical values during calculation. Therefore, in this book, when the word *"pound"* is used for mass, we always refer to it as *"pound mass"* and express it as lb_m. When the word "pound" is used as a unit for the force, we will call it *"pound force"* and express it as lb_f. Now we can define one *pound force*, 1 lb_f, as follows:

$$1\,lb_f = (1\,lb_m)(32.2\,ft/s^2) = 32.2\,lb_m\,ft/s^2 \qquad (3.10)$$

If we try to explain this definition based on Equation (3.3), then one *pound force* (lb_f) is equal to the gravitational pull by the Earth to an object with one pound mass placed on the Earth's surface. However, because the value of g varies at different surface locations on the Earth, we need to be more precise to define one pound force as the gravitational pull by the Earth for an object with one pound mass placed at sea level and at a latitude of $45°$. Of course, we are not restricted to that particular spot on the Earth to use the unit of lb_f. We need to standardize the definition of lb_f strictly based on the numerical values of Equation (3.10). One should notice that this definition is not very convenient when a numerical number other than the unity is used, unlike the SI unit system. In addition, if we do not distinguish lb_f from lb_m, it could become very confusing when one tries to use this definition correctly. To emulate the convenience of the SI system, a new definition was proposed to convert Equation (3.10) to

$$1\,lb_f = (1\,slug)(1\,ft/s^2) \qquad (3.11)$$

and

$$1\,slug = 32.2\,lb_m \qquad (3.12)$$

Instead of one equation, we now need two equations to define the pound force, lb_f, and a new mass unit as the slug is defined by Equation (3.12). In fact, it is easier to stay with Equation (3.10); hence, the unit of slug is not used very commonly.

It is easier to remember Equation (3.10) by stating "One pound force is the gravitational pull at an object with one pound mass placed on (a special spot of) the Earth."

Based on Equation (3.10), we can also establish a unity equation similar to Equation (3.8),

$$1 = \frac{lb_f}{(32.2\,lb_m\,ft/s^2)} = \frac{(32.2\,lb_m\,ft/s^2)}{lb_f} \qquad (3.13)$$

Equation (3.13) is useful to ensure correct unit and calculation when the "American" system of units is used. For example, if an object with one *pound mass* is undergoing 1 ft/s^2 acceleration, then the net force applied to the object is

$$\vec{F_1} = (1\text{lb}_\text{m})\left(\frac{1\,\text{ft}}{\text{s}^2}\right)$$

$$= 1\left(\text{lb}_\text{m}\frac{\text{ft}}{\text{s}^2}\right) \times 1 = 1\left(\text{lb}_\text{m}\frac{\text{ft}}{\text{s}^2}\right) \times \frac{\text{lb}_\text{f}}{\left(32.2\,\text{lb}_\text{m}\frac{\text{ft}}{\text{s}^2}\right)} = \frac{1}{32.2}\,\text{lb}_\text{f} \quad (3.14)$$

The common wrong answer for $\vec{F_1}$ in the above equation is $1\,\text{lb}_\text{f}$, which is off by 3,220%. Major disaster can happen due to such a silly but easy mistake.

Equations (3.8) and (3.13) are useful for converting the force unit from N to lb_f, or vice versa.

Before doing so, we need to define more unity equations.

$$1 = \frac{\text{ft}}{0.3048\,\text{m}} = \frac{0.3048\,\text{m}}{\text{ft}} = \frac{\text{ft}}{12\,\text{in}} = \frac{\text{in}}{25.4\,\text{mm}} = \frac{\text{m}}{1000\,\text{mm}}$$

$$= \frac{\text{mile}}{5280\,\text{ft}} = \frac{0.4536\,\text{kg}}{\text{lb}_\text{m}} \quad (3.15)$$

The conversion between lb_f and N can be determined as

$$1\,\text{lbf} = 32.2\,\text{lb}_\text{m}\frac{\text{ft}}{\text{s}^2} = 32.2\,\text{lb}_\text{m}\frac{\text{ft}}{\text{s}^2} \times \frac{0.4536\,\text{kg}}{\text{lb}_\text{m}} \times \frac{0.3048\,\text{m}}{\text{ft}}$$

$$= 32.2 \times 0.4536 \times 0.3048\,\text{kg}\frac{\text{m}}{\text{s}^2} = 4.451\,\text{kg}\frac{\text{m}}{\text{s}^2} = 4.451\,\text{N} \quad (3.16)$$

However, this result is different from the standard conversion number as $1\,\text{lbf} = 4.448\,\text{N}$ because of the rounded-off numbers used in Equation (3.15). To get $1\,\text{lbf} = 4.448\,\text{N}$, we have to use $1\,\text{lbf} = 32.174\,\text{lb}_\text{m}\,\text{ft/s}^2$ and $1\,\text{lb}_\text{m} = 0.45359243\,\text{kg}$, which yields $1\,\text{lbf} = 4.44821549\,\text{N}$.

By using Equation (3.15) for conversions, we could have 0.0674% error, which may not seem huge but could sometimes be critical. Therefore, it might be useful to establish one more unity equation for convenience

$$1 = \frac{\text{lbf}}{4.448\,\text{N}} = \frac{4.448\,\text{N}}{\text{lbf}} \quad (3.17)$$

Finally, as an exercise, let us calculate how much force is needed to make a typical full size car with $3{,}500\,\mathrm{lb}_m$ achieve $0.5\,\mathrm{g}$ acceleration.

$$\vec{F}_w = (3500\ \mathrm{lb_m})(0.5)\left(32.2\frac{\mathrm{ft}}{\mathrm{s}^2}\right)$$

$$= 56350\left(\mathrm{lb_m}\frac{\mathrm{ft}}{\mathrm{s}^2}\right) = 56350.0\left(\mathrm{lb_m}\frac{\mathrm{ft}}{\mathrm{s}^2}\right)\frac{\mathrm{lb_f}}{\left(32.2\,\mathrm{lb_m}\frac{\mathrm{ft}}{\mathrm{s}^2}\right)}$$

$$= 1750.0\,\mathrm{lb_f} = 1750.0\,\mathrm{lb_f}\times\frac{4.448\,\mathrm{N}}{\mathrm{lb_f}} = 7784.0\,\mathrm{N} \tag{3.18}$$

We have one last remark on the force unit. Quite often, when one refers to the body weight, we will use kg as the unit. This could cause confusion, and we will address this issue when we actually discuss the definition of weight in Section 3.4.

3.3 How to Measure Forces?

Typically, we use the effects of the force to measure it. The effects and the force should be related by fundamental laws. The most important law related to forces, of course, is *Newton's Second Law*, $\sum \vec{F} = m\vec{a}$. This equation appears to be simple but in fact is not very easy to implement. First, the mass of an object, on which several forces, $\sum \vec{F}$, are acting, is generally unknown. Second, the measurement of the acceleration, $\vec{a}$, is not trivial. However, if there is a special case, when $\vec{a} = 0$, *Newton's Second Law* is reduced to a much simpler form,

$$\sum \vec{F} = 0 \tag{3.19}$$

When the force condition satisfies Equation (3.19), we state that the object, upon which the forces are acting, is at a force equilibrium or static equilibrium. We will consider a special case when only two forces are involved in Equation (3.19). If we have a force to be measured, we can try to produce a second force that makes the object stay at static equilibrium. As a result, these two forces must be equal in size, opposite in directions, and at the same point of application. The effects of these two equal-sized forces, such as the induced deformation of the object, can be measured to indicate the size of the forces. This is the basic idea of force measurement using static equilibrium.

However, to have deformations, the object on which the forces are acting cannot be a rigid body. This brings in the fundamental questions discussed in Chapter 2 regarding how we can add forces together on a non-rigid body to achieve the mathematical operation of $\sum \vec{F}$.

As discussed in Chapter 2, we can add forces together as vectors only if they are acting on the same point of application. We can move a force along its line of action only if the object is a rigid body. In addition, only if the object is a rigid body we can move a force out of its line of action to a different line of action (with a resulting moment) to enable an interaction of lines of action, so that forces can move along the lines of action to the same point of application for vector addition.

How can we actually add forces together to satisfy Equation (3.19) on a non-rigid body? Let us reconsider the reason why we cannot add the forces together if they are not acting on the same point on a non-rigid body. It is because the local effects (local deformation) cannot be moved with the force when we move the force along its line of action, as shown in Fig. 2.7 of Chapter 2. Let us again consider two forces acting along the same line of action, but at different points, on a non-rigid body, as shown in Fig. 3.3.

We cannot move the forces along the line of action as in the case of a rigid body because the deformation zone will be different; therefore,

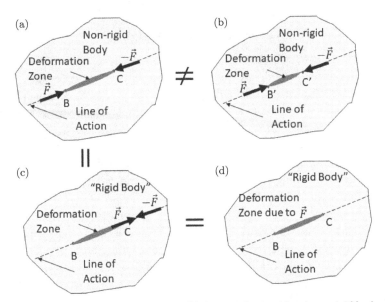

Figure 3.3:   A non-rigid body at force equilibrium can be considered as a rigid body with the deformation preserved and unchanged as in (a), (c), and (d). Case (b) is different from case (a).

Fig. 3.3(a) is not equal to Fig. 3.3(b). However, we recognize that, at equilibrium, the shape of the non-rigid body stays the same after being deformed from the original state. At equilibrium under the same loading condition, the deformed shape of the body will no longer change. As a result, we can consider the deformed shape as a "rigid body" with the deformed shape. We can call it a pseudo-rigid body. We can then move the forces as long as we keep the deformation intact. As shown in Fig. 3.3(c), because the deformation zone is intact, we can move the forces to the same point and eventually cancel them as indicated in Fig. 3.3(d). However, even though we do not show forces in Fig. 3.3(d), the deformation is still there and is recognized as being due to the forces $\vec{F}$ and $-\vec{F}$. The amount of deformation is then a measure to determine the size of $|\vec{F}|$.

Figure 3.3(d) actually expresses a very important concept with practical uses. In general, we do not see forces but only their effects. Therefore, we really don't "see" Fig. 3.3(a), only Fig. 3.3(d). If we can quantify or measure the deformation zone, then we can quantify the forces that cause it. Remember that we do not have the information of motion because $\sum \vec{F} = 0$.

3.4 How Do We Measure Our Body Weight?

In this section, we will show how the static equilibrium condition can be used to measure our body weight. What exactly is the body weight? Most people would think that the body weight is the body mass, which is not correct. Scientifically speaking, when we measure the body weight, we are determining the gravitational pull to our body by the Earth, or the force, $|\vec{F}| = mg$, not m.

Once we find out this gravitational pull, $\vec{F}$, and if we know $\vec{g}$, we can then find out our body mass, m. However, we do not really measure mg directly either, and we will explain why.

As discussed in Section 3.3, we do not really measure a force directly, but its effect. As shown in Fig. 3.4, a simple weight scale concept is illustrated using a spring, which is definitely not a rigid body. Let us assume that the platform on top of the spring is massless for the sake of simpler discussion. We also place a ruler to measure the original length of the spring. We place the Statue of Liberty over the platform and wait until it and the spring reach static equilibrium. We observe that the spring is now compressed, and the amount of deformation can be measured by the ruler. Based on the discussion in the last section, if at equilibrium, the forces acting on the statue must be equal size and at opposite directions; therefore, $|\vec{F'}| = |\vec{F}|$. Based on *Newton's third law*, the force acting on the spring will be $\vec{F'}$ and if the spring is also at equilibrium, $|\vec{F'}| = |\vec{F''}|$. As a result, the deformation of the spring can be used to indicate the gravitational pull on the statue by the Earth, $\vec{F} = m\vec{g}$. This seems to be simple, but the key is that the equilibrium condition must be maintained. We all know that when you step on a scale, you have to remain motionless; otherwise, the weight could fluctuate and be read wrong.

Note that we do not measure the mass, but the gravitational pull, the value of which depends on the value of $\vec{g}$. Therefore, if you move the scale from Paris to the Equator, the weight measurement will be reduced by about 0.25%, as discussed in Section 3.1. This certainly is not a very cost-effective weight loss strategy. On the other hand, if you move the scale to the Moon, your weight will become about 1/6th of that on the Earth because the gravitational acceleration on the Moon is about 1/6th that of the Earth according to Equation (3.4).

Again, when we use a scale to measure the body weight, we do not measure the gravitational pull directly but the forces acting on the spring to cause deformation. For an ideal spring, it is considered as massless. As a result, $|\vec{F'}| = |\vec{F''}|$ must be maintained or the spring will have an acceleration which is infinite. Of course, a real spring is not massless, but this condition of $|\vec{F'}| = |\vec{F''}|$ is still valid, at least for most engineering purposes as long as the mass of the spring is not substantial. Most practical force sensors, such as a spring, rely on this principle without the requirement of static equilibrium. Therefore, the sensing element typically needs to be lightweight or the measurement error could be too big to ignore when the static equilibrium is not achieved. Nevertheless, the equilibrium condition must be maintained for the body on the scale. If the body is not at equilibrium, the force measured by the spring is $\vec{F'}$, not $\vec{F}$ because $\vec{F} - \vec{F'} \neq 0$. The discussion above confirms the statement that we do not measure mg directly.

3.5 Force, Moment Sensors, and Practical Considerations

The simple concept of force measurement device described in Fig. 3.4 can be expanded to use different non-rigid body for achieving deformation. It could be a flexible beam or force-sensitive crystals, such as those used in advanced dynamometers. We can also similarly measure moments.

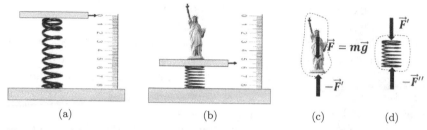

Figure 3.4: (a) A simple weight scale example; (b) weight scale to measure the weight of a statue; (c) FBD for the statue; (d) FBD for the spring.

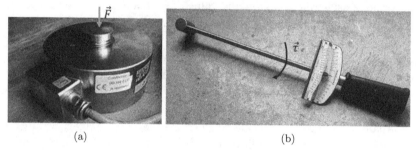

Figure 3.5: (a) A load cell for measuring force and (b) a torque wrench to measure the torque applied.

Source: (a) https://cs.wikipedia.org/wiki/Silom%C4%9Br; (b) https://commons.wiki media.org/wiki/File:Western_Forge_Craftsman_beam_torque_wrench.jpg.

Again, note that the actual sensing element, such as the spring in Fig. 3.4, can still achieve sufficient accuracy without the requirement of static equilibrium of the sensing element as long as the sensing element has a negligible mass. Figure 3.5 is a picture of a commercial force sensor, called load cell, and a tool with a moment sensor, called torque wrench. The load cell converts the force into an electronic signal, while the torque is displayed by an indicator. When we use the torque wrench to apply a torque to tighten a bolt, the tightening action should be slow and firm, maintaining a quasi-static equilibrium condition to get a correct torque reading. If one jerks the handle, the torque reading could be momentarily high, but its reading is not accurate due to the fact that the static equilibrium is not maintained and the torque wrench itself is not negligible in its mass (actually its moment of inertia).

Using the static equilibrium condition to measure a force on a non-rigid body is only one of the many methods for force measurement. The static equilibrium requirement does impose a constraint for the condition when a force could be measured. If we can measure the local effect directly, then a force can be measured without the requirement of static equilibrium. For example, the depression of the right cheek shown in Fig. 2.5 can be used to determine the force acting on the right cheek. The practical difficulty is that we might not have sensors that can be deployed at every location to measure these local effects.

Finally, a force can also be measured dynamically based on *Newton's Second Law* if we can measure the acceleration and combine it with information on the mass. The sensors to measure acceleration are called accelerometers, and they are very commonly used in cars for activating airbags when collisions happen.

3.6 Force and Weight Units Revisited

After the discussion on how to measure body weight using the equilibrium condition, we need to clarify a few confusing usages of weight units. When we use the pound force to describe the body weight, it is naturally related to the gravitational pull by the Earth because of the definition of lb_f. As a result, when we refer to a person who is 150 pounds in weight (i.e. 150 lb_f) on the Earth, the person happens to have a body mass of 150 lb_m. However, when we refer to a person's weight as 70 kg, this actually is incorrect use of engineering units. The unit, kilogram, is the unit for mass, not for force. Therefore, we should say that a person with a 70-kg body mass has a weight as 686.7 N. Of course, we do not state it this way. We simply state that the person weighs 70 kg for convenience.

To avoid confusion, it is better to stay with proper units for mass and force while doing engineering analysis. We should always define the amount of mass, not the weight of a body.

3.7 Concluding Remarks

In this chapter, we discussed the very important concept of force equilibrium or static equilibrium. Once the equilibrium is achieved, the object becomes a pseudo-rigid body, and the forces can be moved along its line of action. We can then perform force additions to compute the combined effect. The force equilibrium is also used to measure forces, such as the gravitational pull to the body. Of course, we need to be careful with the difference between the body weight and the body mass.

The introduction of unity equations is very useful for unit conversions. We must be careful to distinguish between pound mass and pound force to avoid a mistake by 3220%.

Finally, the *Law of Gravitation* was very successful in describing the gravitation pulls between planets, the Sun, and moons of our solar system, until Einstein discovered the Theory of Relativity to point out some major flaws in the *Law of Gravitation*. Of course, this is way beyond the scope of this book, or the author's skillsets.

Chapter 4

How is a Force Applied to an Object?

From the discussions in Chapters 2 and 3, we have obtained proper understanding of forces, their effects, and how to process them for their joined effects using mathematical tools such as the vector analysis. Two very important concepts discussed are related to the rigid body assumption and the pseudo-rigid body assumption when an object is at equilibrium with its shape staying the same after deformation. These two conditions allow us to combine the effects of forces acting at different locations of an object. However, we do not need these conditions to determine what forces are acting on an object.

In this chapter, we explore the answer to the question "How a force is applied to an object?" Knowing the answer to this question, we can independently determine where, which direction, and what size of a force is applied to an object. Applying forces and moments to make products or to achieve desired effects is the objective of most engineering devices.

Recalling from Section 2.1, Chapter 2, we have two different types of forces, contact and non-contact transmitted forces. For the non-contact transmitted force, gravitational force is the most important for most engineering problems in mechanical systems. We also learned about *Newton's Law of Gravitation* in Chapter 3. We conveniently described the gravitational force as a force acting on a specific point of an object. However, if we think a bit deeper, we will be confronted with the question, "Which point?" and "Why only one point?" We will try to answer these two questions in the next section. From here, we will assume an object as a rigid body unless otherwise stated.

4.1 Non-Contact Transmitted Forces

An object with a finite size is not a point. It has a volume, and within the volume, there are different particles distributed at different locations. If the distribution is uniform, we call the object as homogeneous. Let us consider an object with a simple 1D geometry on the Earth, as shown in Fig. 4.1. Of course, there is no such thing as a 1D object, but because the length of this object is so much larger than those in the other two directions, we will call it a 1D object.

In Fig. 4.1(a), we assume this 1D object is formed by many small particles with different masses, m_i. When we place this 1D object on a gravitational field, each particle is subjected to a gravitational pull proportional to the mass. Now, reduce these particles to atomic sizes — there are so many small particles and the forces on these tiny particles can no longer be individually identified. In such a case, we state that the gravity now applies a distributed force on the 1D object as in Fig. 4.1(b). The distributed force is defined as $\vec{f}(x)$, with a unit of N/m. The total amount of the distributed force becomes

$$|\vec{F}| = \int_0^d f(x)dx \tag{4.1}$$

Now, let us assume the object as a rigid body, then we can move every little force, $\vec{f}(x)dx$, to a specific point out of its original line of action, so

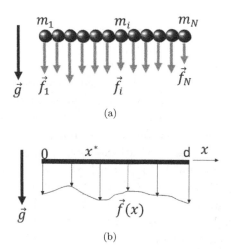

(a)

(b)

Figure 4.1: A 1D object subjected to gravitational forces: (a) many concentrated gravitational forces and (b) a distributed force.

that we can add them together. As discussed in Chapter 2, when we move a force out of its line of action on a rigid body, we need to combine it with a resulting moment. If we want to move every little force to x^* position, the total resulting moment will be

$$(\vec{M})_{x^*} = \int_0^d f(x)(x - x^*)dx \qquad (4.2)$$

Now, an interesting question is "Is there a specific position x^* to render $(\vec{M})_{x^*}$ zero?" The answer is yes, and we simply let $(\vec{M})_{x^*}$ be zero for Equation (4.2) and solve for x^*. For simple distribution of $f(x)$, the solution is readily found. For example, if $f(x)$ is a constant, then $x^* = d/2$. This special point, upon which all the distributed forces by gravity can be represented by a total concentrated force without a resulting moment, is called *Mass Center, or Center of Mass*. We can extend this approach to find the mass center on 1D objects to 2D objects and 3D objects by simply applying Equations (4.1) and (4.2) to the other two directions. For a homogeneous object, the mass center can be defined by its geometric shape. We can easily see that for a homogeneous circular disk, the mass center will be at its center, so also for a sphere. Readers should refer to a standard textbook of Statics or Dynamics for a chart that describes the mass center for different geometric shapes. There are several working problems presented in Chapter 11 related to the calculation of the mass center, such as Problems 11.10–11.12, 11.15–11.16, 11.19–11.21, and 11.24–11.25. We typically use letter G to represent the mass center.

Note that the mass center is independent from the gravitation field. The mass center remains the same if you move an object from the Earth to the Moon.

If we are dealing with discrete particles, Equation (4.2) can be written in a simpler form,

$$(\vec{M})_G = 0 = \sum_1^N m_i(x_i - x_G) \qquad (4.3)$$

where x_G is the x position of the mass center.

For example, consider a simple two-particle system of Fig. 4.2. One particle has a mass of $2m$, while the other has a mass of m. Applying Equation (4.3), the mass center location x_G is found to be $d/3$.

Again, if an object is a rigid body, then all the gravitational distributed forces can be moved to the mass center as one single concentrated force and

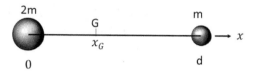

Figure 4.2:   A two-particle system and the location of its mass center.

(a)   (b)

Figure 4.3:   (a) A tennis racket will topple if not supported not at the mass center and (b) it stays leveled if supported at the mass center.

there is no moment. Therefore, theoretically, if we can apply a force on the same location, in the opposite direction, and with the same magnitude, then both forces can be canceled to reach an equilibrium. This observation can be used to find the mass center of an object with a non-standard geometry. For example, if we want to find out the mass center of a tennis racket along its length direction, we can place the racket over a sharp support and the racket will stay horizontal if we support it at the mass center. This is described in Fig. 4.3.

Note that the mass center G of a tennis racket is typically not at the middle point along the length direction. The racket in Fig. 4.3 is considered as head-heavy because the mass center is more toward the head, good for baseliners who like to hit powerful shots behind the baseline. If the mass center is more toward the handle, it is head-light, good for volleys at the net. We use this technique to check the blade of a lawn mower to see if the mass center is at its geometric center. If not, we use a file to take out some excessive mass to achieve it. This is called balancing. Balancing a mower blade is important so that the mower can operate smoothly without excessive vibrations.

However, for a non-rigid body, the mass center location will change when the object changes its shape under loading. In addition, the equivalent concentrated force at the mass center cannot represent the effects of the distributed force. Otherwise, for example, we will not have tidal waves. Tides are due to the gravitational pull from the Moon on the ocean water, which is apparently not at the mass center (roughly the center of the Earth).

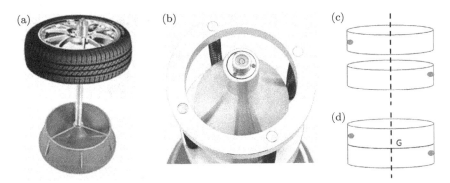

Figure 4.4: A static wheel/tire balancer with a bull's eye level to show if it is leveled. (a) the static balancer with a tire; (b) the bull's level of the balancer; (c) two thin disks with different unbalanced mass; (d) a static balanced thick disk but not dynamically balanced.
Source: (a–b) https://www.ebay.com/itm/Pro-Portable-Hubs-Wheel-Balancer-W-Bubble-Level-Heavy-Duty-Rim-Tire-Cars-Truck-/232408577651.

If we ignore local deformation effects, the total gravitational force acting on the mass center of the Earth (or the Moon) is still useful to describe its orbit. If we extend the idea of Fig. 4.3 to a plate, then we can have a needle point placed at the mass center, and the whole plate should stay leveled. We will need a sensor to tell us that it is leveled with sufficient accuracy. This can be applied to tire balancing to ensure that the mass center of a tire/wheel matches its geometric center using a static balancer (see Fig. 4.4). The bull's eye level is used to indicate if the tire/wheel is leveled horizontally. If not, some additional weight would be added to the rim to bring the tire/wheel back to level. This static balancer typically does not work well because the tire/wheel assembly is not a 2D object. As shown in Fig. 4.4(c–d), if we split a tire/wheel assembly into two disks, the top disk is uniform except for an extra mass on the left side, while the lower disk has the same extra mass on the right side. Combined, the entire assembly is statically balanced with the mass center at the disk center of the disk assembly. However, dynamically, if this disk assembly is under high-speed rotations, the effects of these two extra masses will not cancel each other. To achieve true balancing, an extra mass should be added to each disk to make its mass center at the center. This is the basis of dynamic balancing. This problem is beyond the scope of Statics, but it is worth mentioning here.

Using a needle to balance a 2D object for finding the mass center is not very convenient, and it could take a long time to find the mass center. Practically, if we have a disk with an arbitrary shape that is also

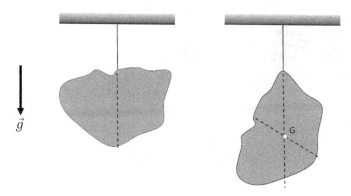

Figure 4.5:   A practical method to determine the mass center of a thin disk.

non-homogeneous, how do we find the mass center? We need to go back to the same concept that the total gravitational force will pass through the mass center. As shown in Fig. 4.5, if we hang the disk by a wire at a point to reach equilibrium, the total gravitational force and the force by the wire must cancel each other. The wire will be vertical and the gravitational force is also vertical. Therefore, the vertical line passing through the wire must be the line of action, through the mass center. Now, we hang the disk at a different location and draw another line of action. The intersection point must be the mass center.

4.2 Contact Transmitted Forces

In the last section, we learned that forces can be applied to an object by another object due *to Newton's Law of Gravitation* without direct contacts. However, most other forces can only be applied through direct contacts or bonds between two objects. Let us consider the case in Fig. 4.6(a). A force, $\vec{R}$, is applied to a homogeneous and rigid block A with mass M, resting on a surface, tied to a rope. Let us just assume that this force, $\vec{R}$, is applied in a certain way without worrying about how it is applied. After the force $\vec{R}$ is applied, we notice that the rope is tightened, but the block stays motionless, which means that equilibrium is achieved. The rope is tightened; therefore, there should be a force applied by the rope to the block, which is labeled as $\vec{T}$. Because the block is subjected to the gravitational pull, there are distributed forces, $\vec{m}(x)$. At the contact surface, there will be frictional forces, $\vec{h}(x)$, distributed along the contact surface, and normal forces, $\vec{n}(x)$, pointing against the contact surface, as shown in Fig. 4.6(b).

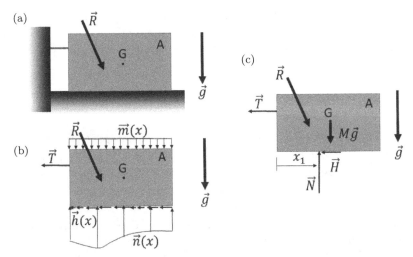

Figure 4.6: (a) An object subjected to a loading condition; (b) both area contact and the gravitational forces are represented by distributed forces; (c) both area contact and gravitational forces represented by concentrated forces.

Because there are many contact points and each point should have its own $\vec{n}$ and $\vec{h}$, they could be different at different contact points. However, since block A is rigid, we can convert the distributed forces into concentrated forces similar to Equation (4.2). As shown in Fig. 4.6(c), the concentrated frictional force, $\vec{H}$, is along the surface and can be anywhere along its line of action. However, the concentrated force, $\vec{N}$, must be placed at a specific point based on Equation (4.2), labeled as x_1, which conceptually is the 'mass center' of $\vec{n}(x)$. Once converted to Fig. 4.6(c), it becomes fairly simple to combine these forces by moving them to a point, for which the mass center of block A is always a good choice.

Let us examine again these forces acting on the block. There is a non-contact force, $M\vec{g}$, acting at the mass center; two forces, $\vec{H}$ and $\vec{N}$, acting on the contact surface; a force, $\vec{T}$, due to the constraint by the rope; and finally a force, $\vec{R}$, which is applied by some means, not specified. In fact, as we will explain further in the next chapter, these are the only four possibilities of how forces can be applied to an object. Again, forces can only be applied by four different ways: (1) by some way, not specifically specified, (2) by non-contact gravitational field (or other fields such as electric and magnetic fields), (3) through contact points, and (4) through constraint points. We will denote theses type of forces as *Applied Forces, Body Forces, Contact Forces, and Constraint Forces*. The names designated here will be discussed in Chapter 5. Note that the "applied forces", the application of which we do not specify exactly, will still be applied through one of the last three ways. Therefore, strictly speaking, there are only three ways a force can be applied to an object.

4.3 Frictional Forces as Contact Forces

When forces are transmitted through a contact surface, there are forces in different directions: one is in the direction normal to the surface, and others are in the directions tangential to the surface. We call the latter as tangential forces. The normal force is easy to understand because it is directly against the surface, and we can easily "feel" it. There are many reasons why there are tangential forces, such as adhesion and friction. We will focus on the frictional phenomena in this section.

Figure 4.7 illustrates a pair of mating surfaces, their contact point, and the normal and tangential directions, line $n-n$ and line $t-t$, respectively. If at the contact point, the velocities in the normal direction are the same for both surfaces, the mating surfaces will stay in contact. Along the tangential direction, if the velocities in the tangential direction are the same, then the two mating surfaces are not slipping with respect to each other; otherwise, they are slipping with respect to each other. Slipping or not, the frictional force can be involved.

Substantial research and engineering efforts are devoted to understand the frictional phenomena, to reduce the frictional effect or to harness

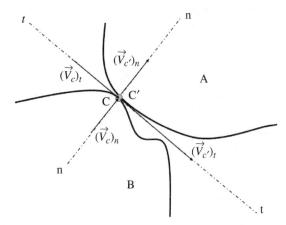

Figure 4.7:   A pair of contact surfaces and their relative velocities at the contact point.

frictional forces for applications. In many examples or homework problems in standard textbooks of Statics, the contact surfaces are assumed to be frictionless, which is rarely the case in practical engineering problems.

Most textbooks on Statics have excellent descriptions of frictional forces. We will not attempt to recite those discussions, but list some key points. In general, the frictional force between two mating surfaces can be categorized into (1) dry friction, (2) viscous friction, and (3) rolling friction. These friction phenomena could occur at the same time.

4.3.1 Dry friction

When two mating surfaces are in contact, a tangential force can develop when there is a relative movement or a tendency of movement between the two surfaces along the tangential direction. As shown in Fig. 4.8(a), this tangential force, denoted as dry friction, can be considered as the interference due to surface roughness. The peaks of surface A can impede the movements of the peaks of surface B. The dry friction can be reduced if the surface roughness is reduced. However, even between two surfaces with mirror finish, frictional force can still develop. This could be attributed to the attractive forces between atoms of the surfaces. Different materials when mating together could have different dry frictional characteristics.

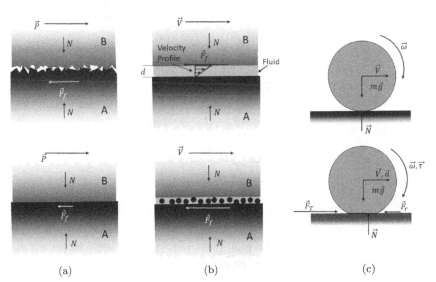

Figure 4.8: Different types of friction: (a) dry friction, (b) viscous friction, and (c) rolling friction.

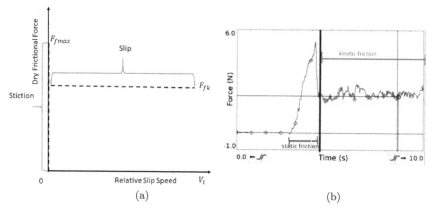

Figure 4.9: (a) Ideal dry friction model and (b) dry friction experimental data from footnote 1.
Source: (b) https://upload.wikimedia.org/wikipedia/en/7/7e/Static_kinetic_friction_vs_time.png.

In dry friction, the frictional force can exist even if the two surfaces are not slipping against each other. If a force, $\vec{P}$, is applied to surface B with surface A stationary, a frictional force, $\vec{F}_f$, will develop to counter this external force. When $\vec{F}_f$ reaches its maximum, it can no longer counter force $\vec{P}$, and so surface B starts to slip with respect to surface A. As the slip occurs, the magnitude of $\vec{F}_f$ drops below the maximum and more or less stays constant independent of the slip speed. This phenomenon was first confirmed by experiments of Coulomb in 1781; thus, it is denoted as Coulomb friction.

Ideally, the Coulomb friction can be represented by a model shown in Fig. 4.9(a), compared with the actual data measured (see footnote 1). The Coulomb friction model is represented by the following equations:

$$F_f < F_{f\max} = \mu_s N, \quad \text{when} \quad V_t = 0$$

$$F_f = F_{fk} = \mu_k N, \quad \text{when} \quad V_t > 0 \tag{4.4}$$

where N is the normal force between the mating surface, μ_s is the static friction coefficient, μ_k is the kinetic friction coefficient, and V_t is the relative slip speed along the tangential line. Note that in the static friction region, the model is not an equation but an inequality. The actual value of the static

---

[1] https://en.wikipedia.org/wiki/Friction.

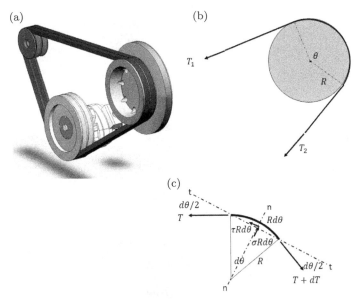

Figure 4.10:   (a) Belt and pulley assembly; (b) belt tensions; (c) infinitesimal belt tension model.
Source: (a) https://upload.wikimedia.org/wikipedia/commons/4/42/Keilriemen-V-Belt.png.

friction can be any value below the maximum static friction $F_{f\max}$, which is related to the static friction coefficient and the normal force. The static friction coefficient, μ_s, is obtained by experiments, as shown in Fig. 4.9(b). Also, as indicated by Fig. 4.9, typically, $\mu_s > \mu_k$.

In general, for machine components, we rarely allow a surface to slip over another surface as dry friction if friction is not desirable. Lubricant will be added between the mating surfaces so that we have viscous friction. In some cases, we rely on dry friction to transmit force (or power) from one component to another. For example, in a belt and pulley assembly, shown in Fig. 4.10(a), we use the static friction to transmit power from one pulley to another.

As shown in Fig. 4.10(b), the power that can be transmitted is $(T_2 - T_1)R\omega$, where T_2 is the tension on the driving side, T_1 is the tension on the driven side, and ω is the rotational speed of the pulley. We can derive a model to show how the belt tension changes from T_1 to T_2.

In Fig. 4.10(c), the forces on an infinitesimal piece of the belt are shown. If we write down the force equilibrium along the $t-t$ and $n-n$ directions, we have,

$$-T \cos\left(\frac{d\theta}{2}\right) + (T + dT) \cos\left(\frac{d\theta}{2}\right) - \tau R d\theta = 0$$

$$-T \sin\left(\frac{d\theta}{2}\right) - (T + dT) \sin\left(\frac{d\theta}{2}\right) + \sigma R d\theta = 0 \qquad (4.5)$$

where τ is the distributed friction stress and σ is the distributed normal stress. As discussed in Section 1.2.4, $d\theta$ is very small. We have $\sin\left(\frac{d\theta}{2}\right) \approx \frac{d\theta}{2}$ and $\cos\left(\frac{d\theta}{2}\right) \approx 1$. Equation (4.5) can be simplified as

$$dT = \tau R d\theta$$

$$T = \sigma R \qquad (4.6)$$

Assuming maximal static friction, or striction, we have $\tau = \mu_s \sigma$. Now, we can combine the above two equations and obtain

$$dT/T = \mu_s d\theta \qquad (4.7)$$

Finally, we have

$$T_2 = T_1 e^{\mu_s \theta} \qquad (4.8)$$

Equation (4.8) indicates that the belt tension can grow exponentially with the angle. In western cowboys movies, a cowboy would wrap the horse's rein around a post a few times and the horse would not be able to escape, due to the power of Equation (4.8). In fact, the friction coefficient in Equation (4.8) can be any value lower than the static friction coefficient for a lower loading condition.

Let us do a simple calculation to show how this can happen. Assume that $T_1 = 1$ lbf, $\mu_s = 0.5$, and $\theta = 4\pi$, then $T_2 = 1 \, e^{0.5 \cdot 4\pi} = 535$ lbf. This is a huge amount of force that can cause great pain to the horse if it tries to pull away.

Another application of dry friction is for the clutch packs of transmissions, as shown in Fig. 4.11(a). Through the dry stiction friction, two plates can be engaged and rotate as one unit. However, if the loading is too high, the clutch packs can slip, entering the kinetic friction region. In a rotor brake system, Figs. 4.11(b–c), we use kinetic friction to dissipate energy into heat in order to slow down a car. For car braking systems, this energy is large, generating a great amount of heat and causing wear of the rotor and the pads.

Figure 4.11: (a) Clutches using stiction dry friction and (b) brakes using kinetic friction. *Source*: (a) https://upload.wikimedia.org/wikipedia/commons/8/86/Basket_clutch.png; (b) https://glossarissimo.files.wordpress.com/2017/03/replacement-rotors-and-brake-p ads.jpg?W=1100; (c) https://c1.staticflickr.com/7/6116/6367413171_446210ac4f_b.jpg.

4.3.2 Viscous friction

As pointed out above, rarely do we allow two mating surfaces to engage with dry friction if friction is not desirable. Lubricant is added to reduce the dry friction. Lubricant could be a thin oil film or a thick layer of fluid (top, Fig. 4.8(b)). Sometimes, solid lubricant can be used, such as fine particles of Molybdenum disulfide (MOS_2) (bottom, Fig. 4.8(b)). In this case, the friction is a combination of viscous friction and rolling friction (discussed in the next section). In viscous friction, the magnitude of the friction is related to the viscosity of the fluid, the slip speed, and the gap between the mating surfaces,

$$F_f = \frac{\mu_v V}{d} A \qquad (4.9)$$

where V is the slip speed, μ_v is the viscosity of the fluid, A is the surface area of the mating surface, and d is the gap between the mating surfaces.

There are many applications when a layer of fluid is used, under pressure, to ensure separation between the mating surfaces, such as journal bearings used in combustion engines and gears of transmissions. The unit of viscosity is $mP_a \cdot s$, or *centipoises*. The fluid viscosity is highly sensitive to temperature; therefore, the viscosity is almost always specified along with temperature. For car engine oils, viscosity levels are specified by Automotive Lubricant Viscosity Grades (SAE Viscosity Grades). For example, a 10W40 engine oil means that the oil at a temperature at $-25°C$ is $7000\,mP_a \cdot s$ during cold cranking (the label of 10W) and $2.9\,mP_a \cdot s$ a at $100\,°C$. For an engine with a 90 mm bore diameter and 90 mm stroke, with 10W40 at $-25°C$, and a piston speed of $3\frac{m}{s}$, during cold cranking, the viscous frictional force is about 60 N for one cylinder. During cruising, at 100C, and a piston speed of 9 m/s, the viscous frictional force is only 0.75 N based on Equation (4.9).

Note that most friction conditions are combinations of dry friction and viscous friction when lubricants are used to reduce dry friction.

4.3.3 Rolling Friction

Rolling friction is a totally different phenomenon. We all understand why tires are used on cars. Interestingly, it is widely believed that native Americans did not use wheels when Christopher Columbus arrived in America, perhaps because Native Americans did not build roads.

As shown in top image of Fig. 4.8(c), there is no friction for a free rolling perfect circular uniform disk without slip on a flat surface. Ideally, this disk can roll forever at a constant speed if the same condition is maintained. There is a subtle observation here. We stated that the disk is rolling without slipping, which means that the velocity of the disk at the contact point is zero, same as the mating point on the ground surface. One may wonder how no slip can occur if there is no friction. Remember that for the Coulomb friction, in the stiction region, the friction could be any value from zero to the static friction maximum. In this case, stiction occurs at the contact point and the friction happens to be zero. If the same condition is applied for a uniform disk rolling down a slope with angle θ with respect to the horizontal plane, there is now a static frictional force against the direction of the disk, as long as the friction does not exceed the maximal static friction. For the same slope, one could find that a block with the same mass could slide down the slope faster than a disk with the same mass. Of course, this occurs if the block is sliding down on a frictionless slope, while the disk is rolling down without slipping. In other words, downhill skiing is faster than downhill skating. This is a topic of Dynamics, and we will not elaborate on it here. We only recognize that when a disk is rolling without slipping, there is a stiction at the contact point, just like the stiction region in dry friction.

In reality, this is certainly not true. One key reason is that the contact condition is not one point but a contact area. The rolling without slipping condition, therefore, is not maintained for the entire contact area. In this case, we state that there exist micro-slips and only one small area where the no-slip condition is maintained. For example, in the case of a real car tire rolling on a flat ground, there is kinetic friction at the contact area, and the concentrated normal force at the contact area is not at the center if the tire is accelerating or decelerating (bottom, Fig. 4.8(c)).

From the above discussions, we recognize that rolling motion can experience a low friction if we can make the contact area approach a contact point (or line). Therefore, it is a common advice to keep your car tires inflated at proper pressure to improve the gas mileage. A harder material can be used for the wheel, such as in the locomotive, as can ball or roller bearings. Ball and roller bearings are also called anti-friction bearings. They are not anti-friction but rolling friction. The key issue to keep a ball/roller bearing running properly and efficiently is the application of preload to make sure that the no-slip condition is maintained. In this way, the friction experienced is in the stiction region. Readers interested in bearing preloads can read Chapter 10 and the articles[2] referenced in this chapter for more details.

Now, let us consider the stiction during rolling as the form of the tractive force or braking force in cars. For a car to accelerate and decelerate, the force comes from the ground, as shown at the bottom of Fig. 4.8(c). The tire will allow the highest acceleration if the entire tire is not slipping (micro-slip is allowed) or skidding. This is because maximum static friction can be achieved in the stiction region. Once the entire tire is slipping, the friction is reduced to the kinetic friction, which would be lower, and the

---

[2]Stein, J.L. and Tu, J.F. (1994). "A state-space model for monitoring thermally induced preload in anti-friction spindle bearings of high-speed machine tools". *Transactions American Society of Mechanical Engineers Journal of Dynamic Systems, Measurements, and Control*, 116(3): 372–386; Tu, J.F. (1995). "Thermoelastic instability monitoring for preventing spindle bearing seizure". *Tribology Transactions*, 38(1): 11–18; Tu, J.F. and Stein, J.L. (1995). "On-line preload monitoring for anti-friction spindle bearings of high-speed machine tools". *Transactions American Society of Mechanical Engineers Journal of Dynamic Systems, Measurements, and Control*, 117(1): 43–53.

tire wear would be higher. Similarly, for a sprinter to win a race, the shoes must not slip on the ground; thus, the use of spikes. The same argument also applies to braking. In order to brake most effectively, the tire must not slip. The tire slip happens when a tire is locked up by the brake pads. Without locking up, the maximal braking force can be achieved at the maximal static friction. Once locked up, the friction is reduced to kinetic friction, and a tire mark can be seen on the ground when braking becomes less effective. As a result, modern cars are equipped with anti-lock braking systems (ABS) to prevent tire locking-up. However, one needs to be careful not to think that ABS will enhance the braking effectiveness. ABS only allows the car to approach the maximal static friction allowed by the ground. If one tries to brake on ice, with a static friction coefficient near zero, ABS will not stop the car at all.

4.4 Force Analysis at Equilibrium

If we choose an arbitrary point and combine all the forces acting on a rigid body, the total resulting force, $\sum \vec{F}$, and total resulting moment, $\sum \vec{\tau}$, must be zero if the body is at equilibrium. Therefore,

$$\sum \vec{F} = 0$$

$$\sum \vec{\tau} = 0 \tag{4.10}$$

We denote these equations as equilibrium-governing equations. For a 2D problem, these represent three equations

$$\sum F_x = 0$$

$$\sum F_y = 0$$

$$\left(\sum M_z\right)_C = 0 \tag{4.11}$$

where point C is any convenient point on the 2D rigid body.

For block A of Fig. 4.6(d), we have four unknowns $\vec{T}$, $\vec{N}$, $\vec{H}$, and x_1, assuming that $\vec{R}$ is given and M is known. As a result, we cannot calculate these forces by solving only three equilibrium equations (Equation (4.11)). It is not uncommon that we have more unknowns and are unable to find them by solving the equilibrium-governing equations. Despite the fact that we cannot readily find out all these unknowns by solving corresponding Equation (4.11), we do know block A is at equilibrium. From an engineering point of view, we can resolve this issue in different ways, as discussed in Chapter 1. First, we can install some sensors to measure some of the forces. For example, if we measure how much the rope is stretched, we can find out $\vec{T}$. The three unknowns, $\vec{N}$, $\vec{H}$, and x_1, can now be determined by solving the equilibrium equations. Another solution is that we can remove the rope so that we have one less unknown. However, this means that we will be relying on the friction to achieve the equilibrium. Finally, we can make the friction between the block and the surface very small via polishing the surface and adding lubricants, then the frictional force, $\vec{F}$, can be assumed to be zero. We can then solve for $\vec{N}$, $\vec{T}$, and x_1. It should be noted that we should not automatically assume that the friction is negligible. We should always assume that there is friction and conduct a thorough analysis to justify if it can be neglected.

For 3D objects, the equilibrium-governing equations have six equations, which means that we can have up to six unknowns

$$\sum F_x = 0$$

$$\sum F_y = 0$$

$$\sum F_z = 0$$

$$\left(\sum M_x\right)_A = 0$$

$$\left(\sum M_y\right)_B = 0$$

$$\left(\sum M_z\right)_C = 0 \qquad (4.12)$$

where points A, B, and C can be any arbitrary points on the rigid body, but they can be the same point, too. Typically, we will choose a point where many forces pass through to simplify the moment equation. These forces result in no moment with respect to this particular point.

4.5 Concluding Remarks

Before we conclude this chapter, it should be noted that we did not introduce any systematic way to determine the contact forces in Fig. 4.6 or forces due to bonding conditions. We did not elaborate enough on the equilibrium condition either. All these issues will be addressed in Chapter 5, which provides a general but systematic method to conduct force analysis for any rigid body under any force conditions.

To recap, knowing how a force is applied to an object allows us to locate the force. It must be at the place where we want to apply, at the mass center, at the contact points, or at the constraint points. These forces are denoted as *Applied Forces, Body Forces, Contact Forces, and Constraint Forces.*

Chapter 5

How to Conduct Force Analysis Correctly Most of the Time?

The title of this chapter is a bit conservative. As engineers, we should strive for correct force analysis all the time, not just most of the time. A simple force analysis test is given in the first lecture for the engineering courses taught by the author as a survey for force analysis skills. The correct rate of force analysis at best has been 10%, be it sophomores, seniors, or graduate courses. Most of the force analyses appear more like random guess than trained engineering analysis. Why so? All these students have completed Engineering Statics before coming to my classes. The reason that I found is related to how they conduct the force analysis. Mostly, it was done in an arbitrary way. Therefore, the success of their force analysis is based on luck. We cannot design engineering products by chance. We must seek 100% success, aside from occasional slips due to unforeseen mistakes. We should not count on genius or unique talent to achieve this success either. Therefore, what is the best way to achieve correct force analysis results? The answer is that we must be thorough and systematic. To be thorough, we have to be exhaustive, seeking out all possibilities. By doing force analysis systematically, we can avoid mistakes due to carelessness and omissions. Force analysis is not a creative process, but proper exercise of rigorous engineering disciplines.

How can we be thorough? We must consider all the possible forces that could act on the object. As discussed in Sections 4.1 and 4.2, we pointed out that all possible forces acting on an object can be grouped into four types: *Applied Forces, Body Forces, Contact Forces*, and *Constraint Forces*. We will present a systematic way to account for all these four different force types to avoid any omissions. Remember, if you miss a force, the entire force analysis is wrong.

We will discuss these four force types further in the following sections. Before we do so, we need to discuss the concept of the free-body diagram, which is the most common method for force analysis. A free-body idea is about isolating a portion of a system for which we want to conduct a thorough force analysis.

5.1 Free-Body Diagram

We will not address exactly how this term, "free-body", was proposed. Instead, we can interpret it literally that we want to "free" a body from the rest of the world. However, we must do this without changing the force condition that the object is subject to. In our engineering world, we hardly have a body which is free, isolated, or as an ideal point with an infinitesimal size. On the other hand, we cannot afford to do a force analysis for the entire world all the time. Therefore, we should only consider a specific "body" part of an object to conduct the force analysis. To do so, we will "free" or isolate this "body" of interest.

5.2 Draw a Boundary Around the Free-Body

To isolate this "body" of interest, we must identify it first. The best way to identify it is to draw a boundary around it. In this way, we can clearly indicate what this "free-body" is, what are included within this "free-body", and how it is connected to the rest of the world. For example, if we are interested in designing a better backpack for reducing lower back pain (Fig. 5.1(a)). The first intuition would be to isolate the backpack and consider the forces acting on the backpack as shown in Fig. 5.1(b). However, the objective is to design a backpack to reduce the lower back pain. Therefore, it is probably better to isolate a free-body which includes the upper body and the backpack, as shown in Fig. 5.1(c). The boundary line cutting through the lower back carries the most significance because this part of the boundary is where the contact forces or constraint forces will reside. We will elaborate this further in the next sections. Lower back pain must be related to the forces acting at the lower back, mainly on the spinal vertebrae. We can extend the boundary to the knees if we are interested in the burden to the knees. The decision of choosing a free-body by drawing

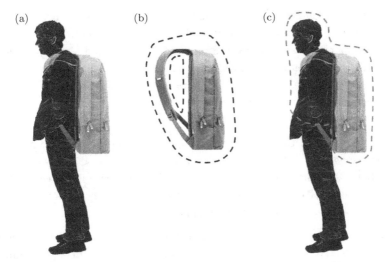

Figure 5.1:   (a) A person with a backpack; (b) FBD for the backpack; (c) FBD for the upper body and the backpack.

a boundary around it is not trivial. One would become more efficient and choose a "better" free-body when he or she becomes more experienced.

5.3 The ABCC Method for Force Accounting

Once we draw a boundary to isolate a free-body for analysis, the next step naturally is to consider all possible forces that could act on the free-body. As we discussed before, there are only four different types of forces: *Applied Forces, Body Forces, Contact Forces*, and *Constraint Forces*. Taking the first letter of these four types of forces, we have ABCC. Each of these four letters represents a force type. If we go through this list of ABCC, we will exhaust all possible forces, and avoid missing any forces. These four letters are easy to remember, as easy as ABCC.

Most mistakes students would make in the free-body diagram analysis is due to omitting forces. They simply mark down some forces here and there, but they rarely are confident that they have included all possible forces. By going through this list of ABCC, such a mistake can be avoided.

Now let us elaborate on these four force types further.

Applied forces (A): Applied forces are those forces we want to apply to the free-body. We do not have to specify how an applied force is applied (contact or non-contact), but we need to specify where, which direction, and how much the force is. For homework problems, a statement such as, "There is a force of 500 N applied to the top center of the block, vertically down", indicates that an applied force should be included. An applied force could also be a loading condition for which we would like to verify if it could cause damages. In conclusion, applied forces are those forces we need to consider but which are not specifically related to body forces, contact forces, and constraint forces.

Body forces (B): Body forces are non-contact forces due to gravitational, electric, magnetic fields, etc. As discussed in Chapter 4, body forces act on a body as distributed forces. For a rigid body, the magnitude of the gravitational body force is mg, its direction is along the gravitational acceleration, $\vec{g}$, and it acts at the mass center.

Contact forces (C): Contact forces, represented by the first letter **C**, are the first type of contact transmitted forces. They occur at the contact points where the free-body is in touch with the rest of the world. A contact is not permanent, and the free-body can be readily separated at the contact point. For example, when your feet are on the floor, there are contact areas between your shoes and the ground. You can readily lift your feet up and lose the contact. When we put our hands together, palm against palm, we create a contact. The contact is only created when two hands are pressed against each other, but we lose the contact when we pull them apart. This is an important observation because the force perpendicular (or normal) to the contact surface is always compressive. Whenever there is a contact, there are contact forces. If we draw a boundary through a contact, we must replace the contact by corresponding contact forces at the boundary. A ball that rests on a surface has a contact point, but it is not permanently bonded to the surface. When we only have a contact point, we have concentrated contact forces. If the contact is a line or an area, we have distributed contact forces. However, similar to body forces, it is easier to replace the distributed forces as concentrated forces at specific locations, as discussed in Chapter 4. Sometimes, it is important to determine the distribution of the contact forces. In such cases, the contact forces should be represented by the profile of the distributed forces.

Constraint forces (C): Constraint forces, represented by the second letter **C**, are the second type of contact transmitted forces. Constraint forces are different from the first type of contact transmitted forces. Constraint forces occur when we try to isolate a free-body by "cutting" it off the rest of the world. For example, if we want to make a hand a free-body, then we have

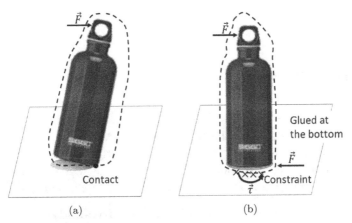

Figure 5.2:   (a) A water bottle in contact with a surface and (b) a water bottle glued to a surface.

Source: https://marchantscience.wikispaces.com/envisrh.

to "cut" the hand off at the wrist. Our hand and the arm are not just in contact. We can pull our hand hard, but the hand won't lose the "contact" with the arm. This is an important distinction of a constraint from a contact. Furthermore, when we isolate a free-body through a constraint point, line, or area, we might need to replace the constraint not only with forces but also moments. This difference is further illustrated in Fig. 5.2.

As shown in Fig. 5.2, when a water bottle is placed upon a surface, there is a contact between the bottom of the water bottle and the surface. When a force is applied to topple the water bottle, the contact can only offer a limited counter-moment to resist the toppling, as shown in Fig. 5.2(a). On the other hand, if the water bottle is glued to the surface, the constraint by the glue can offer a strong counter-moment to resist the toppling. We do not have to label the counter-moment explicitly if we specify an asymmetric distributed force across the contact or constraint area.

However, we should not conclude that a contact can never offer a counter-moment. For example, for the same water bottle in Fig. 5.2, if we spin the water bottler around the vertical axis, the friction at the bottom of the water bottle can still resist the spin. This counter-moment is limited in the toppling case, because the water bottle can lose the contact and the normal contact force can only be compressive.

Finally, the forces and moments related to contacts and constraints are all located at the boundary. Therefore, we cannot emphasize enough how important it is to draw a boundary as the first step for the free-body diagram analysis.

5.4 The ABCC Method in Three Steps

We propose a systematic and thorough procedure for the free-body diagram analysis and denote it as the ABCC method. Let us use the case in Fig. 5.3 to demonstrate how to use the ABCC method. A force, $\vec{R}$, is applied to a homogeneous and rigid block A with mass M, resting on a surface, tied to a rope. After the force $\vec{R}$ is applied, we notice that the rope is tightened, but the block stays motionless, which means that an equilibrium is achieved.

Step 1: Draw the boundary to isolate the free-body

The choice of boundary is not trivial as discussed before. For this case, if we are interested in verifying if the rope is strong enough to keep the block from moving, we should draw the boundary through the rope. We then draw around the entire block to form a complete loop, or a closed boundary, as shown in Step 1 of Fig. 5.3.

Step 2: Go around the boundary to mark all the contacts and constraints

Once the boundary is drawn, we should start at any point of the boundary and go around the boundary to mark all the contacts and constraints. Use

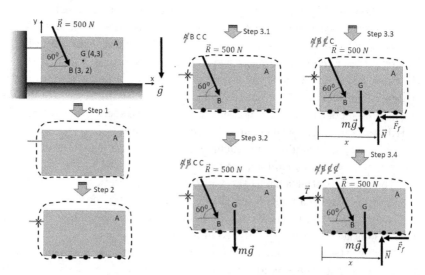

Figure 5.3:   Steps involved to conduct a thorough and systematic force analysis.

a heavy dot to represent a contact point, or several heavy dots for a contact line or area, as shown in Step 2 of Fig. 5.3. Use a cross to represent a constraint. In this case, the constraint is at the rope where the boundary cuts through.

Step 3: Account for all the forces simply as ABCC

Now, we are ready to account for all possible forces by following through the ABCC list. Write the four letters, ABCC, on the side of the free-body.

Step 3.1: A for applied forces

Ask yourself, "Do we have any applied forces?" In this case, the answer is "yes" because there is an applied force, $\vec{R}$, specified in the problem statement. Draw it and mark it correctly onto the free-body diagram, as shown in Step 3.1 of Fig. 5.3. Cross out the letter A from the ABCC list to indicate that all applied forces have been accounted for.

Step 3.2: B for body forces

Now, ask yourself, "Do we need to consider body forces?" The answer is "yes" in this case because the block is in a gravitational field. We can identify the mass center and mark a concentrated gravitational force, $m\vec{g}$. Typically, it is better to draw the gravitational force from the mass center G and downward, as shown in Step 3.2 of Fig. 5.3. Cross out the letter B to indicate that the body forces have been accounted for.

Step 3.3: C for contact forces

Now, ask yourself, "Do we have any **contact** points?" Look at the boundary to check if there are any heavy dots (which we drew in Step 2). In this case, we have many dots to form a contact line under the block. Instead of marking distributed forces across the contact line, we can just mark a concentrated normal force, $\vec{N}$, and define an application point, which is x distance away from the left edge. The normal contact force can only be compressive; therefore, the force must be pointing toward the boundary. Next, we mark the concentrated frictional force, $\vec{F}_f$. We only need to draw the frictional force in the most probable direction. If the assumed direction is wrong, the force value will be negative once we solve the force equations. Sometimes, it is critical that we draw the frictional force direction correctly. In such cases, we need to judge the tendency of the moment (see Problem 11.26 in Chapter 11). Cross out the first letter C to indicate that all the contact forces have been accounted for, as shown in Step 3.3 of Fig. 5.3.

Step 3.4: C for constraint forces

Finally, ask yourself, "Do we have any **constraint** points?" Look at the boundary again to check for cross marks. In this case, we have a constraint at the rope. The rope can only sustain tensile force; therefore, the constraint force, $\vec{T}$, should be drawn away from the boundary. Cross out the second letter C to indicate that all the constraint forces have been accounted for, as shown in Step 3.4 in Fig. 5.3. Again, tensile forces must be pointing away from the boundary, while the compressive forces, pointing toward the boundary.

Mission accomplished!

Once we have gone through Steps 1–3, we have accounted for all possible forces, thoroughly and systematically. We will be confident that we have constructed a proper free-body diagram for force analysis. The next stage is to examine if we can determine all the forces identified by the free-body diagram. Finally, we do not need to label the forces with the vector signs as long as we mark the direction of the force clearly.

5.5 Force Analysis at Equilibrium

If we choose an arbitrary point and combine all the forces, the total resulting force $\sum \vec{F}$ and moment, $\sum \vec{\tau}$, must be zero if block A is at equilibrium. Therefore,

$$\sum \vec{F} = 0 \tag{5.1}$$

$$\sum \vec{\tau} = 0 \tag{5.2}$$

We denote these equations as the force equilibrium equations. For a 2D problem, these represent three scalar equations. The equilibrium equations with respect to mass center G become

$$R\cos(60°) - T - F_f = 0 \tag{5.3}$$

$$-R\sin(60°) - mg + N = 0 \tag{5.4}$$

$$\vec{r}_{B/G} \times \vec{R} + 2T\,\vec{k} - (x - 4)N\,\vec{k} - 3\,F_f\vec{k} = 0 \tag{5.5}$$

where $\vec{r}_{B/G} = \vec{r}_B - \vec{r}_G = -1\,\vec{i} - 1\vec{j}$ and $\vec{R} = 500\,(-\cos(60°)\vec{i} - \sin(60°)\vec{j})$.

For Equations (5.3–5.5), we have four unknowns T, N, F_f, and x_1, but only three equations. As a result, we cannot solve for all the unknowns. This is not uncommon. Despite the fact that we cannot readily find out these unknowns, block A still reaches an equilibrium. As discussed in Chapter 1, we can resolve this issue in several ways. First, we can assume that the friction is zero. By assuming the friction as zero, we are determining the tension of the rope as the worst-case scenario. Without friction, the rope needs to carry all the loading due to $\vec{R}$, according to Equation (5.3). Therefore, a rope size chosen by this assumption will be able to handle the worst and maximal loading condition. If the objective is not about choosing a proper size for the rope, but precise force calculations, we can install sensors to measure some of the forces. For example, if we measure how much the rope is stretched, we can determine T. The unknowns are reduced to three, N, F_f, and x_1. They can now be determined by solving the force equilibrium equations. Another solution is that we can remove the rope. However, we will be counting on the friction to achieve equilibrium. Finally, we can make the friction between the block and the surface very small via polishing the surface and adding lubricants. As a result, the frictional force, F_f, can be assumed to be zero. We can then solve for T, N, and x_1. This is how the worst-case scenario described above is satisfied. It should be noted, yet again, that we should not automatically assume that the friction is negligible.

For 3D objects, the equilibrium force equations will have six equations, which means that we can have up to six unknowns.

5.6 Backpack Example Revisited

Let us revisit the example of Fig. 5.1 and apply the ABCC method. Because the objective is to design a better backpack for reducing lower back strain, we choose the boundary as shown in Step 1 of Fig. 5.4. The entire backpack is inside the boundary. In Step 2, we go around the boundary and mark all the contact and constraint points. The contact points where the boundary passes through are at the contact between the backpack and the butt of the person. The constraint points are at the boundary where it cuts through the lower back. Remember that these contact and constraint points are only on the boundary. Therefore, we do not consider the contact between the back and the backpack, or the contact between the top strap and the shoulder. These contact points are inside the boundary. In Step 3, we go through the list of ABCC.

Step 3.1: We do not have any Applied forces. Cross the letter A out.

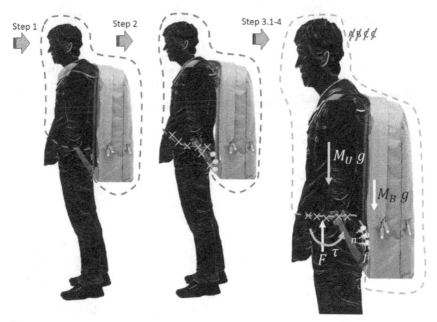

Figure 5.4:   Free-body diagram construction using the ABCC method for the backpack analysis.

Step 3.2: For Body forces, we have two masses to consider. One is the upper body of the person. We mark $M_U g$ for the gravitational force due to the upper body mass, M_U. We draw a downward arrow from approximately where the mass center of the upper body is. The other mass is the backpack. We label it as $M_B g$ and mark it accordingly at the mass center of the backpack. Cross the letter B out.

Step 3.3: For Contact forces, we note that we have some contact points (dots). We mark the normal forces, n, and frictional forces, f, and draw the arrows accordingly. Cross the first letter C out.

Step 3.4: For Constraint forces, we note that we have constraint points (crosses). We mark a concentrated force, F, and a moment, τ, at the constraint points across the lower back. Cross out the second letter C.

Again, Mission Accomplished! Correct and complete!

By examing this free-body diagram in Fig. 5.4, our goal is to reduce the force, F, and the moment, τ, which cause lower back strain. Without solving the force equilibrium equations, we can quickly see that if we reduce our upper body weight (i.e. $M_U g$), or carry less inside the backpack (i.e. $M_B g$), it will help. Also, a thinner backpack to keep $M_B g$ closer horizontally to the lower back will reduce the moment τ. Finally, we also notice that the contact forces between the backpack and the butt are actually counteracting the gravitational forces of the upper body and the backpack. This observation leads to a possible backpack design to shift some weight of the backpack to below the lower back so that the force and the moment to the lower back are less. Finally, if we can have a design so that there is a force pushing up at the bottom of the backpack (for example someone is holding it), the lower back strain will be greatly reduced. This might lead to a backpack design with some sticks tightened to the lower legs of the person so that the backpack weight is partially carried by the lower legs directly without going through the lower back. Of course, we can also mount a stick with wheels to the ground to carry some weight of the backpack.

5.7 Concluding Remarks

Before we conclude this chapter, it should be noted that we did not introduce a systematic way to determine the forces due to contact and bonding conditions. We did not elaborate enough on the equilibrium condition either. All these issues will be addressed in Chapter 6 to determine the contact and constraint forces systematically for common mechanical devices.

Finally, according to the teaching experience of the author, students often discount the importance of not skipping any steps of the ABCC method. Once taught the ABCC method, a student might think that it is so easy and he or she does not need to go through the entire steps anymore. For example, they will not draw the boundary. Without the boundary, the contact and constraint points are no longer properly defined. Often, they will not actually go through the ABCC list. According to the personal experience of the author, if the author does not follow through every step of the ABCC method, mistakes could still happen. It is, therefore, a puzzle to the author why most students, immediately after being taught the ABCC method, feel so comfortable in not following it.

Chapter 6

How Do We Set a Free-Body Free?

In fact, a free-body is not free. A free-body is still subjected to many forces and moments, but it is now free of physical contacts and constraints with the rest of the world. In other words, when we construct a free-body diagram, we are replacing the contacts and constraints by equivalent forces and moments. It would be helpful if we have a database to show how to carry out this replacement for different contacts and constraints often found in practical mechanical systems.

In standard textbooks of statics, there are tables or charts to show how to carry out this replacement. However, these charts are not organized on the basis of how forces are transmitted through contacts and constraints. As a result, they are not very easy to follow and are difficult to implement correctly.

In this chapter, we will provide detailed discussions and charts, organized according to the ABCC method, i.e. contact conditions first, and then constraint conditions, going from 1D and 2D to 3D situations. Practical examples are also provided so that readers can identify contacts and constraints correctly for actual mechanical systems.

6.1 Common Contact Conditions

As defined in Chapter 5, contact points are not permanent bonds. A free-body can be readily separated from the neighboring objects at these contact points. When two objects are in contact, the contact could be a point, a line, or an area. The line could be straight or curved, while the contact area could be a flat or curved surface.

6.1.1 Point contacts

Point contacts between 3D objects: As shown in Fig. 6.1, a point contact is formed between a sphere and a flat surface (Fig. 6.1(a)), between two spheres (Fig. 6.1(b)), between two misaligned cylinders (Fig. 6.1(c)), between a sharp corner of a cube and a flat surface (Fig. 6.1(d)), and between a needle and a flat surface (Fig. 6.1(f)). These contact conditions can be extended from flat surfaces to curved surfaces and from spheres to non-spherical surfaces. Multiple and countable contact points, such as four contact points between a sphere and an internal cone (Fig. 6.1(e)), can also be formed.

Point contacts between 2D objects: As shown in Fig. 6.2, a point contact is formed between a thin disk and a flat surface (Fig. 6.2(a)), between two thin disks (Fig. 6.2(b)), between the sharp corner of a thin plate and a flat surface (Fig. 6.2(c)), between a thin disk and a curved surface (Fig. 6.2(d)), between two curved 2D surface (Fig. 6.2(e)), and between a needle and a flat surface (Fig. 6.2(f)).

In Fig. 6.2, we also define the tangential line (t–t) and the normal line (n–n) at the contact point. At a 2D contact point, the normal contact force will be at the contact point along the normal line, while the frictional force will be along the tangential line (t–t) in either direction depending on the tendency of the movement. For the 3D objects shown in Fig. 6.1, we define two tangential lines and one normal line at the contact points (see Figs. 6.3(b1)–6.3(b3)).

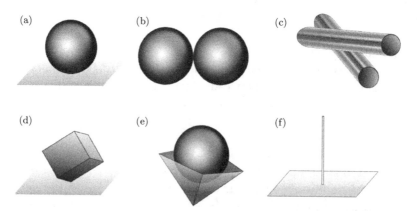

Figure 6.1: 3D objects with point contacts: (a) between a ball and a flat surface; (b) between two balls; (c) between two skewed cylinders; (d) between the corner of a cube and a flat surface; (e) between a ball and the inner surface of a cone; (f) between a needle and a flat surface.

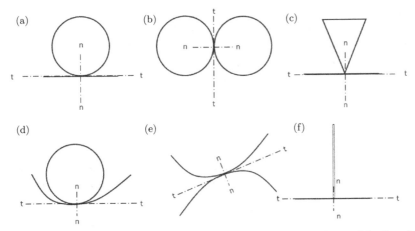

Figure 6.2:   2D objects with point contacts: (a) between a disk and a straight line; (b) between two disks; (c) between a triangle tip and a straight line; (d) between a disk and a curved line; (e) between two curved lines; (f) between a needle and a straight line.

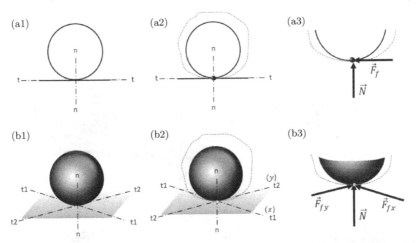

Figure 6.3:   How to replace a contact point to equivalent forces: (a) a 2D case and (b) a 3D case.

6.1.2 Equivalent force condition at the point contact

Based on the point contact condition, if we remove the neighboring object in contact with the free-body, we need to replace this neighboring object with an equivalent normal force and frictional forces. This is shown in Fig. 6.3.

In Figs. 6.3(a3) and 6.3(b3), the normal force, $\vec{N}$, is always compressive and it should be pointing toward the boundary; the frictional force can, however, be in any direction along the tangential line. For a 3D point contact, we can assume a pair of tangential lines perpendicular to each other as an x–y coordinate. As shown in Fig. 6.3(b1), the tangential line pair (t1–t1 and t2–t2) can be rotated to any orientation. For 3D contact points, we have two frictional forces and one compressive normal force, as shown in Fig. 6.3(b3).

6.1.3 Line contacts

As shown in Fig. 6.4, a line contact is formed between a cylinder and a flat surface (Fig. 6.4(a)), between two cylinders (Fig. 6.4(b)), between a

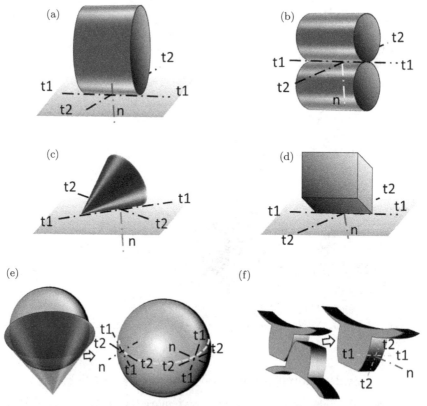

Figure 6.4: 3D objects with line contacts: (a) between a cylinder and a flat surface; (b) between two aligned cylinders; (c) between a cone and a flat surface; (d) between an edge of a cube and a flat surface; (e) between a ball and the inner surface of a cone; (f) between two engaged gears.

cone and a flat surface (Fig. 6.4(c)), between an edge of a cube and a flat surface (Fig. 6.4(d)), between the internal surface of a cone and a sphere (Fig. 6.4(e)), and between two gear teeth (Fig. 6.4(f)). The contact lines can be straight lines (Figs. 6.4(a)–6.4(d) and 6.4(f)), a circular line (Fig. 6.4(e)), or a curve. Similarly, we can define the tangential lines and normal line for the line contact. One of the tangential lines will be along the contact line, while the other will be perpendicular to it. The normal line is then defined accordingly. For a circular contact line, the second tangential line and the normal line are different at different contact points on the contact circle (Fig. 6.4(e)).

6.1.4 Area contacts

As shown in Fig. 6.5, an area contact is formed between a cube and a flat surface (Fig. 6.5(a)), between a cylinder and a ring (Fig. 6.5(b)), between a cone and a tapered ring (Fig. 6.5(c)), between a sphere and an internal sphere in a ball joint (Fig. 6.5(d)), between a diamond and a ring (Fig. 6.5(e)), and between a bolt and a nut (Fig. 6.5(f)). The contact area can be a straight surface (Fig. 6.5(a)), a cylindrical surface (Fig. 6.5(b)), a cone surface (Fig. 6.5(c)), a spherical surface (Fig. 6.5(d)), facets of a

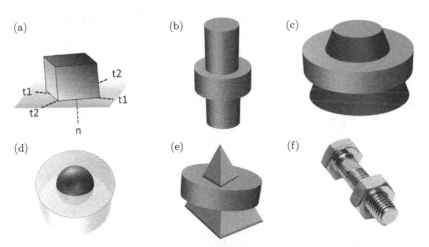

Figure 6.5: Area contact between different 3D objects: (a) between a cube and a flat surface; (b) between a cylinder inside a larger cylinder; (c) between a cone inside an inner cone; (d) between a ball and its housing; (e) between a pyramid and its covering; (f) between the bolt and the nut.

Source: (f) http://pngimg.com/download/3001.

diamond cone (Fig. 6.5(e)), or threaded surface (Fig. 6.5(f)). For Fig. 6.5(a), we can define the tangential lines and the normal line related to the area contact. For other area contact surfaces, the tangential and normal lines are defined locally at different points within the contact surface.

6.1.5 Free-body representation for line and area contacts

Similar to the point contact condition, if we remove the neighboring object in contact with the free-body, we need to replace this neighboring object with equivalent normal and frictional forces for the line or the area contact. This is shown in Fig. 6.6. However, because a contact line or a contact area can have many (uncountable) contact points, we have distributed frictional forces and normal forces.

If we choose to draw the entire distributed force profiles as shown in Fig. 6.6(a3), we will replace the line contact with the frictional distributed force, $\vec{f}$, and the distributed normal force, $\vec{n}$. However, typically, it is very difficult to obtain the correct profile of these distributed forces. For some idealized cases, certain models can be used for the distributed force profile prediction, such as the Hertzian contact theory for contacts between elastic bodies, or sensors can be used to measure the force distribution. As discussed before, we can convert a distributed force into a concentrated force acting on a specific point. For a line contact, we will then replace it with a

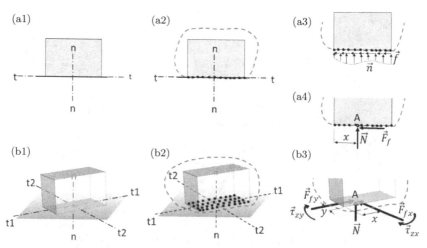

Figure 6.6: Equivalent contact forces for line and area contacts: (a) a 2D case and (b) a 3D case.

concentrated normal force, $\vec{N}$, and a concentrated frictional force, $\vec{F}_f$. The frictional force, $\vec{F}_f$, is along the tangential line, while the normal force, $\vec{N}$, needs to be placed at a specific point, defined by a distance, x, from the left edge of the contact line, shown as point A in Fig. 6.6(a4). It should be noted that the location of the point of application of $\vec{N}$ is very important and must not be ignored. As a result, for a line contact in a 2D case, we have three unknowns, $\vec{F}_f$, $\vec{N}$, and x. For a line contact in a 3D case, we will have three contact forces and two location variables, for a total of five unknowns.

Similarly, for an area contact, if we convert all the distributed forces to concentrated forces, as shown in Fig. 6.6(b3), we have three concentrated forces, two location variables, and two additional moments. The two location variables, x and y, are related to the point of application of the normal force, $\vec{N}$, shown as point A in Fig. 6.6(b3). The line of action of the frictional force, $\vec{F}_{fx}$, however, does not necessarily pass through point A. We can move $\vec{F}_{fx}$ to point A, resulting in a moment, around axis z, denoted as $\vec{\tau}_{zx}$ since it is related to $\vec{F}_{fx}$. Similarly, we have z, $\vec{F}_{fy}$, and $\vec{\tau}_{zy}$. Thus, there are seven unknowns. The number of unknowns presents a difficult problem. In many practical applications, however, the force distribution of the contact area may be symmetric. If so, we can quickly determine the location of point A at the geometric center of the contact area and there will be no moments $\vec{\tau}_{zx}$ and $\vec{\tau}_{zy}$. As a result, there will be only three unknowns, $\vec{N}$, $\vec{F}_{fx}$, and $\vec{F}_{fy}$.

Of course, there are many cases when the moments $\vec{\tau}_{zy}$ and $\vec{\tau}_{zx}$ can be significant. One example is the frictional force between the tire and the road when one makes a turn. There will be a counter-moment, $\vec{\tau}_z$, to partially turn the tire back when one lets go of the steering wheel.

6.1.6 Common parts with point, line, and area contacts

A few common mechanical devices with point, line, or area contacts are shown in Fig. 6.7. For ball bearings (Fig. 6.7(a)), the contact between the ball and the bearing's inner (outer) ring is typically considered as a point contact, or more specifically a 2D point contact because the ball can only move along the circumferential direction. In the case of roller bearings (Fig. 6.7(b)), the contact between the roller and the inner (outer) ring is a line contact. The contact line could be straight or curved. Similarly, this line contact is a 2D line contact. Typically, we can assume symmetric contact forces for the analysis of roller bearings. For an HSK tool holder

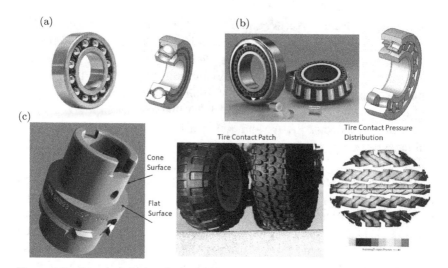

Figure 6.7: Practical devices for different contact conditions: (a) point contact, ball bearings; (b) line contact, roller bearings; (c) area contact, tool holder and tires.

Source: (a) http://www.911uk.com/viewtopic.php?p=1141885, https://commons.w ikimedia.org/wiki/File:Angular-contact-ball-bearing_single-row_din628_type-b_120.png; (b) https://www.thingiverse.com/thing:781635, https://commons.wikimedia.org/wiki/ File:Spherical-roller-bearing_double-row_din635-t2_120.png; (c) https://commons.wiki media.org/wiki/File:DIN_69893_HSK_A63_3drender_1.png, https://www.flickr.com/pho tos/42988571@N08/7524617342, https://commons.wikimedia.org/wiki/File:Tirefootpri nt.jpg.

shown in Fig. 6.7(c), there are two contact areas, one related to the cone and one related to the flat surface of the ring. Conventional 7/24 tool holders may have the cone surface contact only. Note that once the surface contact is locked in by locking devices, such as bolts, nuts, and clamps, the contact becomes a constraint, which we will discuss in the following sections. Another important contact area example is the contact patch between a tire and the road. The force distributions on a tire contact patch are very complicated and have been the focus of intensive research by tire companies so as to enable them to optimize the tire performance based on traction, wear, comfort, stiffness, and energy consumption. The moments due to the area contact in Fig. 6.6(b3) are the phenomena we experience on making a turn while driving. At the end of the turn, if you let go of the steering wheel, the steering wheel will rotate back partially by itself due to the presence of the moments.

6.1.7 Practical consideration of point and line contacts

In practice, there are no true point and line contacts. A point contact is likely a small circular or oval area contact, while a line contact is a narrow rectangular area contact. For example, if we are interested in the motion analysis of a car, we will conveniently assume the contact between a tire and the road as a point contact. However, if we are interested in assessing the tire wear, the contact will be an area contact with highly complicated distributed force profiles. We make these assumptions based on the objective and the precision requirements of the analysis.

If we consider point, line, and area contacts as area contacts, the size of the contact area is the smallest for the "point" contact and the largest for the area contact. For the same force, the contact stress (force/area) is higher for a smaller contact area. The material failure is typically proportional to the magnitude of the contact stress. As a result, the loading capacity, or the amount of the force that a contact area can carry, is the smallest for a point contact and the largest for an area contact. This is very true, as can be seen by the fact that we know how easily a sharp needle can penetrate through the skin with a small force while an unsharpened pencil cannot despite being applied with a much larger force.

On the other hand, a point contact or a line contact is involved with fewer unknowns; therefore, it is easier for analysis and more predictable for the loading condition.

As engineers, we balance between these considerations to choose different contact conditions for our design and analysis.

6.2 Common Constraint Conditions

As defined in Chapter 5, constraint points are permanent bonds unlike contact points. A free-body cannot be readily separated from the neighboring objects at these constraint points. There are many ways two objects can be "bonded" together through constraint points. Those important constraints, which are discussed here, include the following: (1) cable and wire constraint, (2) pin and hinge constraint, (3) ball joint constraint, (4) non-spherical surface constraint, (5) pined–pined rod constraint, (6) sliding collar constraint, and (7) total constraint. We will discuss the features of these constraints and determine their equivalent forces and moments.

6.2.1 Wire/cable/rope constraint

One of the simplest constraints is achieved by using a wire, a cable, or a rope. Wires (or cables or ropes) are flexible. They can only be subjected to tensile force because they buckle or bend when compressive forces are applied. For engineering applications, the mass of a wire is typically much less than that of the object to be constrained by the wire. We can reasonably assume that a piece of wire is massless. As shown in Fig. 6.8(a), a plumb bob is attached to a rope and hung vertically for a vertical reference, commonly used in construction. The general case of a wire constraint is shown in Fig. 6.8(b). If we draw a boundary across the wire to isolate a free-body (Fig. 6.8(c)), we should mark an "×" where the boundary intersects with the wire. We can also isolate a piece of the wire as a free-body as shown in Fig. 6.8(d). We realize that the two forces acting on the wire free-body must be in tension, thus pointing away from the boundary and they must be of an equal size and in opposite directions, aligned with the stretched wire, to reach an equilibrium because the wire is assumed to be massless. Recall that we used the same assumption in Section 3.4 of Chapter 3 for the spring for weight measurement. As a result, based on *Newton's Third Law*, the constraint force at the point where the "×" is marked should be the same as the tensile force as shown in Fig. 6.8(c). The wire constraint is the easiest constraint. One should note that the constraint force at the wire constraint must be tensile, pointing away from the boundary, and aligned with the stretched wire. For this constraint, the only unknown is the magnitude of the force because the direction is determined by the orientation of the wire.

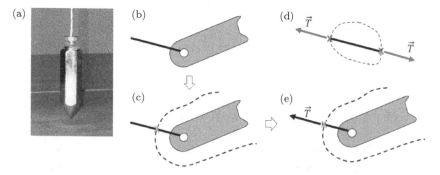

Figure 6.8:   (a) A plumb bob; (b)–(e) the construction of the corresponding FBD. *Source:* (a) https://asacredrebel.wordpress.com/2012/07/15/gods-plumb-line/.

6.2.2 Pin/hinge constraints and their extensions

A pin or a hinge constraint is used to join two objects so that they can be joined at the pin location while being capable of rotating freely with respect to each other, as shown in Fig. 6.9(a). These two objects will be rotating on the same plane (or on parallel planes). Therefore, a pin constraint is considered a 2D constraint, and we can represent it as shown in Fig. 6.9(b), where a linkage is pined to a foundation at point A. The linkage can rotate freely around point A, but point A will be stationary. The pin constraint provides constraint forces to keep point A from moving but provides no constraint moment to restrict the rotation of the linkage around point A. If we want to remove this pin constraint, we should draw a boundary across the pin, as shown in Fig. 6.9(c), and mark an "×" at the center of the pin. The equivalent constraint forces will be two forces $\vec{F}_x$ and $\vec{F}_y$ as shown in Fig. 6.9(d).

These two forces can be combined as one resultant force pointing to a specific direction. The two unknowns become the magnitude and the direction of the combined resultant force. In general, it is easier and more convenient to use $\vec{F}_x$ and $\vec{F}_y$ when we derive the force equilibrium equations. Note that $\vec{F}_x$ and $\vec{F}_y$ are plotted to the right and upward, respectively. However, the actual directions can be reversed. In other words, the combined resultant force can be compressive or tensile (pointing toward or away from the boundary). We simply assume the most probable directions. If the directions are wrong, the magnitudes of the forces will be negative. As a 2D constraint, the pin constraint is the second simplest constraint, and no moment is needed.

In practice, all objects are 3D objects. In a practical pin constraint as shown in Fig. 6.9(a), the linkages can only rotate on an x–y plane. Therefore, a pin constraint, when viewed as a 3D object, can offer moment constraints around both x and y directions as shown in Fig. 6.10, in

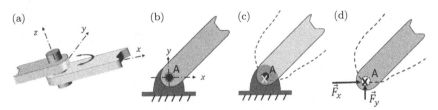

Figure 6.9:   (a) A pin linking two members; (b) a member fixed by a pin; (c–d) construction of the corresponding FBD.

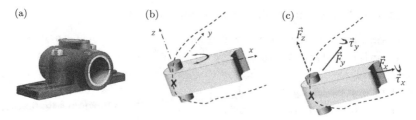

Figure 6.10: (a) A deep flange as a 3D pin; (b) motion of a 3D pin; (c) FBD of a 3D pin.
Source: (a) https://upload.wikimedia.org/wikipedia/commons/thumb/c/c0/Pillow-block-bearing.jpg/1200px-Pillow-block-bearing.jpg.

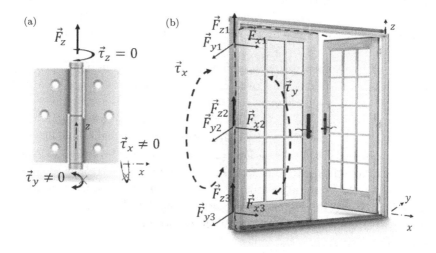

Figure 6.11: (a) The movements and constraints of a hinge and (b) multiple hinges used to secure a door.
Source: (a) https://www.publicdomainpictures.net/en/view-image.php?image-9211&picture=hinge&jazyk=JP; (b) https://www.flickr.com/photos/andersenwindows/6237870220/in/photostream.

particular for a deep flange as shown in Fig. 6.10(a). A 3D pin can also be made to resist the linear motion along the z-axis so that a constraint force $\vec{F}_z$, plus two moments, $\vec{\tau}_x$ and $\vec{\tau}_y$, exist.

However, when designing a mechanical system, we should avoid using the moment constraint capability of a 3D pin for carrying heavy loading. In other words, it is better to design a system with several 2D pins to carry moments. This is the typical case of using hinges to secure a door, as shown in Fig. 6.11. Hinges, as shown in Fig. 6.11, only allow rotation around the

z-axis, and they do not allow rotation around the x and y axes. However, we should not rely on one hinge to secure a door to keep it from rotating around the x and y axes. The moment capacities of a single hinge are too low (Fig. 6.11(a)). To achieve high load-carrying capacity to resist $\vec{\tau}_x$ and $\vec{\tau}_y$, we use multiple hinges at different locations. As shown in Fig. 6.11(b), three hinges are used for each door. Each hinge will only carry forces in x and y directions, as well as the weight of the door in the z direction. The forces in the x directions of these hinges combine to form a moment to counter $\vec{\tau}_y$, while the forces in the y directions form a moment to counter $\vec{\tau}_x$, as shown in Fig. 6.11(b). Just like 2D pins, we only have to replace each hinge with two forces as shown in Fig. 6.9(d).

Another example of using several 2D pin constraints is the universal joint. As shown in Fig. 6.12(b), a shaft is attached to a rod with two pin joints to allow rotation around the x-axis (line AB), while a second shaft is attached to a second rod to allow rotation around the y-axis (line CD). These two rods are then joined together as a rigid cross, as shown in Figs. 6.12(a) and 6.12(b). A universal joint can be considered as four pin constraints put together in a 2D plane. As can be seen from Figs. 6.12(b) and 6.12(c), there are four pins, A, B, C, and D. Each pin will provide two

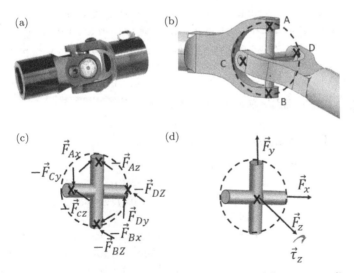

Figure 6.12: (a) A university joint; (b–d) the construction of the corresponding FBDs. *Source*: (a) https://mechanics.stackexchange.com/questions/24550/clicking-noise-in-op el-astra; (b) http://catiatutorial.blogspot.com/2009/11/modeling-universal-joint-in-ca tia-v5.html.

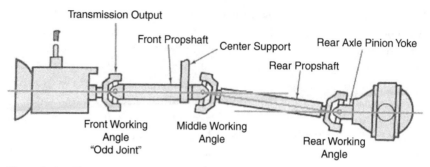

Figure 6.13:   Two universal joints are used to connect two different axes of drive shafts. *Source*: https://mechanics.stackexchange.com/questions/6800/1-piece-versus-2-piece-d rive-shaft.

constraint forces as shown in Fig. 6.12(c) for eight forces, which then form two force couples for those forces in the $x-y$ plane. The combination of these forces and force couples can be summarized into three constraint forces and one moment, as shown in Fig. 6.12(d). Because the cross is rigid, these two rods are not allowed to rotate with respect to each other. As a result, a universal joint can provide constraint forces in all three directions but only a moment constraint around the z-direction, as shown in Fig. 6.12(d). The z-axis is defined as normal to the plane defined by the cross.

A universal joint allows a rotation to be transmitted from one shaft to another shaft that is not aligned in a straight line with the first shaft. Two universal joints are often used for the main drive shaft of a rear-wheel drive car when the engine crank shaft is higher than the shaft to the differential, as shown in Fig. 6.13.

6.2.3 Ball joint constraint

A ball constraint is used to join two objects so that they can be joined at the ball joint location, allowing for free rotation with respect to each other around any axis, as shown in Fig. 6.14(a). A ball joint constraint is considered a 3D constraint.

The linkage connected to the ball joint can rotate freely, but the center of the ball will be stationary. Therefore, the ball joint constraint provides constraint forces in three directions to keep the ball center from moving but provides no constraint moment to restrict the rotation. If we want to remove this ball constraint, we should draw a boundary across the ball center, as shown in Fig. 6.14(c), and mark an "×" at the center of the ball.

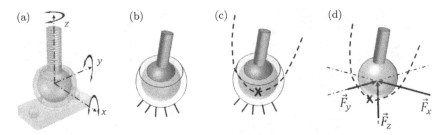

Figure 6.14:   (a) A ball joint; (b–d) the construction of the corresponding FBDs. *Source*: (a) https://www.thingiverse.com/thing:939326.

The equivalent constraint forces will be three forces, $\vec{F}_x$, $\vec{F}_y$, and $\vec{F}_z$, as three unknowns, as shown in Figure 6.14(d).

Ball joint constraints are used in many important applications, such as in the suspension and steering systems of automobiles (Figs. 6.15(a) and 6.15(b)), in metrology for ballbar measurement (Fig. 6.15(c)), and in artificial knee replacement (Fig. 6.15(d)). For the automobile applications, the ball joints are subjected to huge loading. The fit between the ball and its socket must be tight without play, to maintain accurate positioning. Typically, grease is applied to reduce the wear of the ball. Rubber boots are used to keep the grease from being contaminated by dirt. When a rubber boot breaks due to aging, dirt can get into the interface between the ball and the socket, causing severe wear. Excessive ball joint wear in the car's suspension system can cause large play and excessive tire wear. Worst of all, it could lead to catastrophic failures when the ball separates from the socket, leading to major car accidents. Ball joints for automotive applications are not failure-proof designs, and so drivers should pay attention to their conditions.

6.2.4 Non-spherical surface constraints

Figure 6.16 shows the joining between a nut and a wrench as an example of non-spherical constraint. This constraint contains an extra moment, $\vec{\tau}_z$, in addition to two forces $\vec{F}_x$ and $\vec{F}_y$ (Fig. 6.16(c)). However, the two forces are not essential. For the hand tool example, one might think this is a contact condition because the wrench can be easily removed from the nut. However, we should consider it as a constraint.

Non-spherical constraints are typically used for transmitting torques. More examples are shown in Fig. 6.17. The example in Fig. 6.17(a) is a valve

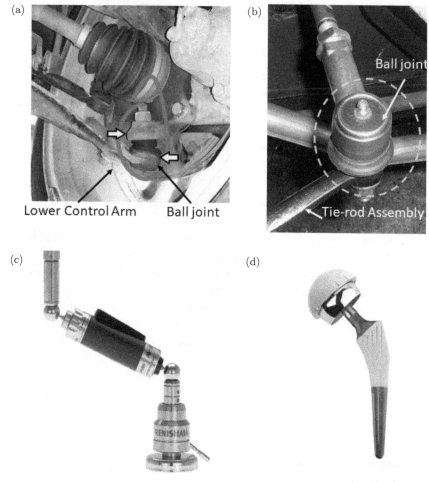

Figure 6.15:   Ball joints used for different applications: (a) lower control arm; (b) tie-rod assembly; (c) ball bar measurement tool; (d) an artificial knee.
Source: (a)   https://mechanics.stackexchange.com/questions/9884/macpherson-strut-lower-control-arm-ball-joint-play-and-suspension-geometry;   (b)   https://www.flickr.com/photos/mwanasimba/16966732482/;   (c)   http://ims-eng.net/products/renishaw/; (d) https://www.geripal.org/2013/01/metal-on-metal-hip-replacements-tragic.html.

handle that fits into a square stem for turning the valve or a faucet open or close. In Fig. 6.17(b), it is a shaft coupling using splines. In Fig. 6.17(c), it is the square fitting for rachet wrenches.

The ability of a non-spherical contact surface to sustain both forces and a torque has many other applications. One popular application is the

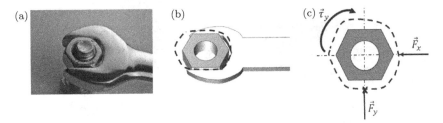

Figure 6.16: Non-spherical constraint examples: (a) a wrench tightening a nut; (b–c) the construction of the corresponding FBDs.
Source: (a) https://imbratisare.blogspot.com/2013/02/el-fmi-sigue-apretando-las-tuer cas-los.html.

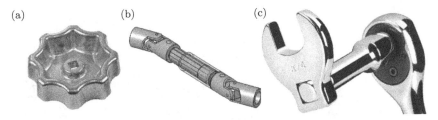

Figure 6.17: Non-spherical constraint examples: (a) a valve handle; (b) a spline coupling; (c) a ratchet wrench and a crowfoot tool.
Source: (a) https://diy.stackexchange.com/questions/125237/what-is-the-missing-part-that-would-fix-this-outdoor-spigot; (b) https://commons.wikimedia.org/wiki/File:Car dan-joint_spline-shaft_3D_transparent_animated.gif; (c) https://diy.stackexchange.com/questions/103422/remove-fastener-nut-for-kitchen-faucet-rusted.

trailer hitch shown in Fig. 6.18. With a lock pin and a square contact surface, this trailer hitch is capable of sustaining forces and moments in all three directions. This is called a full constraint, and it will be discussed in Section 6.2.7.

6.2.5 Pined–pined rod constraints

There is a special case of the pined constraint, which was discussed earlier in Section 6.2.2. As shown in Figs. 6.19(a) and 6.19(b), a pined–pined rod is defined as a 2D element constrained by two pins on each end and no external forces on the element except at the pins. The corresponding free-body diagram is shown in Fig. 6.19(c). Normally, we should label two force components on each pin constraint, but in this special case, we only need to label one force along the rod direction. This is because the element is

(a) (b)

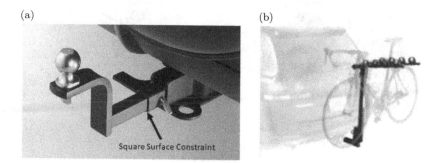

Square Surface Constraint

Figure 6.18:   (a) A trailer hitch and (b) a bicycle rack attached to a trailer hitch. *Source*: (a) https://bicycles.stackexchange.com/questions/8958/bike-rack-for-hatchback -vehicle; (b) http://www.cyclecityusa.com/index.php?title=Accesories.

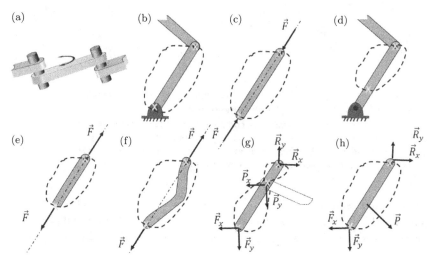

Figure 6.19:   (a) a member pined at both ends only; (b–f) the construction of the corresponding FBDs; (g–h) a member with additional forces or constraints.

at equilibrium, and the only possibility is that the resultant force at each pin must cancel each other to result in this equilibrium. As a result, if we draw a straight line between these two pins, the resultant force at each pin must align with this straight line as shown in Figs. 6.19(c) and 6.19(f). We can draw the boundary across the middle of the beam, and the force at the constraint of the beam (not a pin) will still be aligned with the rod, as shown in Figs. 6.19(d) and 6.19(e). Therefore, a 2D element with only two pined constraints, without any other forces, is similar to a rope constraint, except that the force could be tensile or compressive. A 2D element with the pined–pined rod constraint is very useful because we already know the force direction at the pined constraint. Therefore, we only have one unknown — the magnitude of the force.

It is very important that there are only two pined constraints and no external forces acting on the element other than on the pins. If these conditions are not met, as shown in Figs. 6.19(g) and 6.19(h), then we do not have a pined–pined rod constraint. In Fig. 6.19(g), there is a third pined element in the middle, i.e. more than two pins. Therefore, we need to replace each pined constraint with two unknown forces. In Fig. 6.19(h), there is an external force, $\vec{P}$, acting on the element. As a result, we need to replace each constraint with two unknown forces.

In reality, it is very difficult to satisfy the condition that requires no external forces acting on the element in locations other than those of the two end pins. The gravitational force will always be acting on the mass center of the element. However, if the constraint forces are expected to be substantially larger than the weight of the element, we can ignore the gravitational force.

There are many important applications of the pined–pined rod constraint as shown in Fig. 6.20. In Fig. 6.20(a), a bridge was built with many beams, called trusses, which are riveted only on the end. We will discuss truss design in greater detail in Chapter 7. It is important to know that these trusses are not weightless, but we can assume that the weight is split in two equal parts to act on each end pin. Typically, the force sustained by the truss is much higher than its weight to justify this assumption. Figure 6.20(b) shows a multiple-axis suspension design for automobiles. Each axis is formed by a 2D element with the pined–pined rod constraint. In this way, we know each force is applied (compressive or tensile) along each axis. For a 5-axis suspension design, only one degree of freedom is allowed (up and down motion). Figure 6.20(c) shows a shock absorber (circled element), used in the rear suspension of a bicycle. This pined–pined rod is no longer rigid but allows movement along the longitudinal direction. The movement is resisted by viscous friction inside the rod to dissipate vibration energy. Figure 6.20(d) depicts a hydraulic cylinder (circled element) used in a crane hoist. The hydraulic pressure inside the cylinder extends along its length and can withstand forces along the longitudinal direction.

Figure 6.20: Practical applications of pined-pined rod constraints: (a) a bridge; (b) multiple-axis rear suspension system; (c) shocks of a bicycle; (d) hydraulic actuator of a crane hoist.
Source: (a) http://www.bridgeofweek.com/2009; (b) https://www.deviantart.com/fl -history/art/Lotus-97T-Rear-Suspension-Belgium-1985-384304250; (c) https://en.wiki pedia.org/wiki/File:Mountain_Bike_Suspension.jpg; (d) http://pngimg.com/download/ 20218.

6.2.6 Sliding collar constraints

If we allow a pin to move along a slot, we have the sliding collar constraint. It can also be a collar over a rod as shown in Fig. 6.21(a).

A sliding collar constraint is different from a sliding contact because the normal force between the sliding pin and the slot can be either compressive or tensile.

In fact, the pin is in contact with the slot at two points, but only one of the contact points will be engaged and subjected to compressive normal forces. We do not have to identify which contact point is in contact. We only have to label a normal force and allow this force to be in either of the directions. The frictional force is usually small compared with the normal force, in particular if lubrication is applied. Therefore, we only identify one normal force in a sliding collar constraint, as shown in Figure 6.21(c–d).

Examples of elements with a sliding collar constraint are shown in Fig. 6.22, which includes a power window regulator used in cars

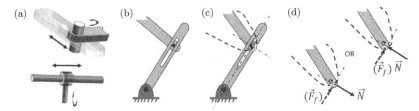

Figure 6.21: (a) sliding collar examples; (b) 2D representation; (c–e) the construction of the corresponding FBDs.

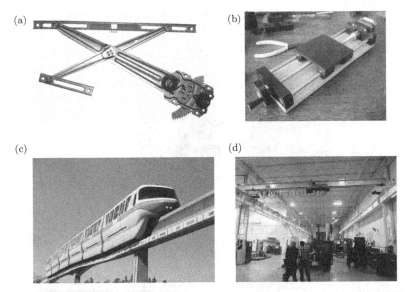

Figure 6.22: Sliding collar examples: (a) an automatic window regulator; (b) a sliding stage; (c) a monorail train; (d) an overhead crane hoist.

Source: (a) https://robotics.stackexchange.com/questions/1935/how-to-open-a-sliding-window; (b) https://www.thingiverse.com/thing:898212; (c) https://en.wikipedia.org/wiki/Walt_Disney_World_Monorail_System; (d) https://www.flickr.com/photos/obrien installations/7703847972.

(Fig. 6.22(a)), guide rails used in a linear stage of a servo control positioning system (Fig. 6.22(b)), a rail for monorail trains in cities (Fig. 6.22(c)), and the overhead beam used for a sliding crane host (Fig. 6.22(d)).

6.2.7 Total constraint

The last constraint type is called total or full constraint. When an element is attached to another element with a full constraint, there will be no

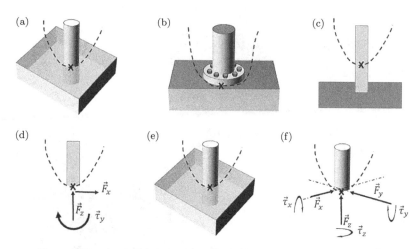

Figure 6.23: (a–c) Total constraints and (d–f) the corresponding FBDs.

Figure 6.24: The total constraint examples: (a) bolted post; (b) bridge foundations; (c) workpiece and tool held by chucks; (d) welded tubes.

Source: (a) http://www.philobiblon.com/nigropress/; (b) https://en.wikipedia.org/wiki/Chaotianmen_Bridge; (c) https://www.flickr.com/photos/designandtechnologydepartment/5162652710; (d) http://newprotest.org/categories.pl?forging.

relative movement between these two elements. A full constraint means that all degrees of freedom have been eliminated between these two elements. As shown in Fig. 6.23(a), a rod is buried inside a plate such that the rod and the plate are bonded together as one element. Similarly, a post can be bolted to a plate to achieve the same constraint (Fig. 6.23(b)). In a 2D case, the full constraint should be replaced with two equivalent forces and a moment as shown in Figs. 6.23(c) and 6.23(d). Note that the boundary can cut across anywhere on the rod, which would result in different forces and moments. Similarly, in a 3D case, the full constraint should be replaced with three equivalent forces and three equivalent moments.

Examples of the full constraints are shown in Fig. 6.24, which includes a post bolted to a plate (Fig. 6.24(a)), the bridge foundation buried firmly into the river bed (Fig. 6.24(b)), a workpiece held tightly by a three-jaw chuck and a centering drill bid held by a drill chuck (Fig. 6.24(c)), and a few square tubes welded together (Fig. 6.24(d)).

6.3 Contact and Constraint Table

Table 6.1 summarizes all the contact and constraint cases discussed in Section 6.2. Students should refer to this when they construct freebody diagrams.

6.4 Concluding Remarks

Table 6.1 provides a convenient reference for constructing free-body diagrams. In the contact cases, the normal forces are always compressive, while in the constraint cases the force and moment can be in either direction. At equilibrium, there are three forces/moments governing equations in a 2D case, while there are six forces/moments governing equations for a 3D case. Therefore, when constructing free-body diagrams, it is important to keep track of unknowns when drawing a boundary to "free" a body. In a 2D case, the unknowns are limited to three, while in a 3D case the unknowns number six. In some special cases, we can add additional governing equations based on geometry or other conditions (see Chapter 8). However, it is better to observe the number of unknowns when constructing a free-body diagram.

Table 6.1: Free-body diagram reference table.

Type/application	Free Body	Equivalent Forces/Moments, 2D	Equivalent Forces/Moments, 3D	Remarks
Point contact				The normal force is always compressive. Friction can be neglected if the surface is smooth or lubricated.
Line contact			—	The equivalent forces can be represented as distributed or concentrated. As a concentrated force, the application point of the normal force should be identified.
Area contact		—		The application point of the normal force should be identified and the two horizontal forces could be moved to the same point with resulting moments.
Wire constraint			—	The force is always tensile, pointing away from the boundary.

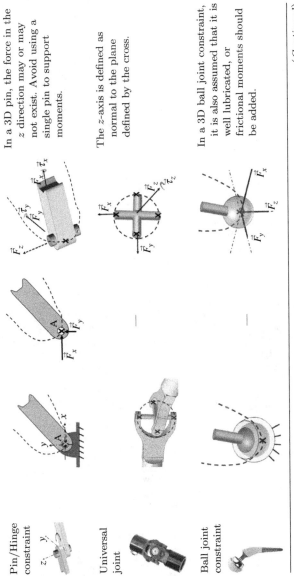

Pin/Hinge constraint

In a 3D pin, the force in the z direction may or may not exist. Avoid using a single pin to support moments.

The z-axis is defined as normal to the plane defined by the cross.

Universal joint

Ball joint constraint

In a 3D ball joint constraint, it is also assumed that it is well lubricated, or frictional moments should be added.

(Continued)

Table 6.1: (*Continued*)

Type/application	Free Body	Equivalent Forces/Moments, 2D	Equivalent Forces/Moments, 3D	Remarks
Non-spherical constraint		$\vec{F}_x$, $\vec{F}_y$, $\vec{\tau}_y$	—	The main purpose of a non-spherical constraint is for applying a torque.
Pined–pined rod constraint		$\vec{F}$	—	The force is aligned with the straight line passing through both pins.
Sliding collar constraint		$\vec{N}$, $(\vec{F}_y)$	—	Frictional force can be neglected if well lubricated. The normal force could be in either direction.
Total constraint		$\vec{F}_x$, $\vec{F}_z$, $\vec{\tau}_y$	$\vec{F}_x$, $\vec{F}_y$, $\vec{F}_z$, $\vec{\tau}_x$, $\vec{\tau}_y$, $\vec{\tau}_z$	The total constraint will have three unknowns in the 2D case and six unknowns in the 3D case.

Chapter 7

Is Free-Body Diagram Analysis
Really Useful?

Conducting correct force analysis is critical in the design stage for many applications. Apart from solving homework problems in standard text-books, can free-body diagram analysis be used for practical applications? The answer, of course, is YES!

In this chapter, we will provide force analysis for several real-world applications using the ABCC method and the contact/constraint chart presented in Chapter 6 for constructing correct free-body diagrams.

7.1 Trusses

7.1.1 Degrees of freedom of rigid body assemblies

A truss system contains straight members pined together only at their ends; therefore, these members are subjected to the pined–pined rod constraint. As discussed in Chapter 6, the forces acting on these truss members are always along the member, either in tension or in compression.

Because a truss system is typically used as a structure to sustain external forces without movement, a truss system has zero degrees of freedom. We will discuss the 2D truss system first. An unconstrained 2D rigid body has three degrees of freedom. We can pick an arbitrary point of a 2D rigid body, and this point can have a translation movement on a plane, while the rest of the body will appear to rotate with respect to this point, as shown in Fig. 7.1(a). Note that a 2D rigid body will have only one rotation because of the rigid body assumption. As a result, this rigid

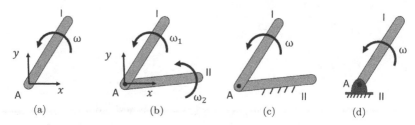

Figure 7.1: The degrees of freedom of 2D assemblies: (a) a free 2D member; (b) two 2D members pinned together; (c) one fixed and one pinned members; (d) a 2D member with a fixed pin.

body has three degrees of freedom, shown as x, y, and ω. If we know the motion of one point and we know the rotation of the rigid body, then we can define the motion of any point on this rigid body.

If we connect a second member (II) to member I via a pin, as shown in Fig. 7.1(b), this assembly now has four degrees of freedom. Both members are free to rotate, but they are joined at point A, with A free to move in the $x - y$ plane. If we fix member II (as shown in Fig. 7.1(c)), it becomes a fixed pin as shown in Fig. 7.1(d). The assembly has only one rotation left. When we see a mechanism similar to that in Fig. 7.1(d), we should remember that it is actually a two-member system as shown in Fig. 7.1(c). Gruebler's equation is often used to calculate the degrees of freedom in a rigid body assembly. Gruebler's equation is written as

$$M = 3\,n - 3j_3 - 2j_2 - 1\,j_1 \qquad (7.1)$$

where M is the degrees of freedom of the assembly, n is the total number of members of the assembly, j_3 is the number of the joint that eliminates all three degrees of freedom, j_2 is the number of the joint that eliminates two degrees of freedom, and j_1 is the number of the joint that eliminates one degree of freedom.

Apparently, a fixed joint as shown in Fig. 7.1(c) to fix member II is a j_3 joint. A fixed joint can also be achieved by bolting, welding, bonding, etc. A pin joint is an example of a j_2 type of joint, which eliminates two translational motions but allows for rotation. There are other types of j_2 joint, to be discussed later along with the j_1 joint. For the assembly shown in Fig. 7.1(d), we only have one pin joint; therefore, the degree of freedom is calculated to be

$$M = 3\,(2) - 3(1) - 2(1) - 1\,(0) = 1 \qquad (7.2)$$

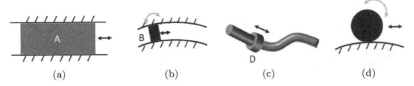

Figure 7.2: Examples of the j_2 joint: (a) a block within a straight grove; (b) a block with a curved groove; (c) a collar over a curved shaft; (d) a disk rolling without slipping over a curved surface.

Let us look at other examples of the j_1 and j_2 joints in 2D assemblies. In Fig. 7.2, examples of the j_2 joint are shown. In Fig. 7.2(a), rigid body A is confined in a straight groove and only allowed to move horizontally without rotation. Therefore, it has only one degree of freedom. The contact condition between rigid body A and the groove is a j_2 joint.

Applying Gruebler's equation, the degree of freedom of rigid body A is calculated to be

$$M = 3\,(2) - 3(1) - 2(1) - 1(0) = 1 \qquad (7.3)$$

Note that the grove could be curved, as shown in Fig. 7.2(b). When rigid body B is moving along the curved groove, it has both horizontal and vertical movements, but the movement path is predefined and rigid body B cannot deviate from this path. Thus, it has only one degree of freedom in translation. Furthermore, even though there is a rotation of rigid body B when it moves along this curved groove, this rotation is not independent from the translation movement, as indicated by its lightly shaded arrow sign. We know exactly the orientation of rigid body B at any specific location along the groove. Therefore, the joint of Fig. 7.2(b) is the same as that of Fig. 7.2(a). Similarly, the j_2 joint could be a collar moving along a rod, straight or curved, as shown in Fig. 7.2(c). In Fig. 7.2(d), a roller is rolling without slipping over a surface (curved or straight). Because there is no slipping, the linear motion of the roller center and the rotation of the roller are related, not independent (thus, the rotation is indicated as a lightly shaded sign). As a result, the roller in Fig. 7.2(d) has only one degree of freedom. A rolling-without-slipping joint is, therefore, a j_2 joint, which eliminates two degrees of freedom.

In Fig. 7.3, examples of the j_1 joint are shown. In Fig. 7.3(a), rigid body A is confined in a straight or a curved groove and is only allowed to move along the groove and rotate at the same time. The contact between rigid body A and the groove is not rolling-without-slipping. Therefore, rigid body

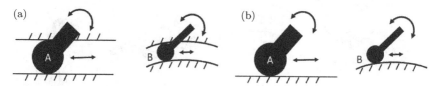

Figure 7.3: Sliding contact as the j_1 joint: (a) a disk sliding inside a groove and (b) a disk sliding over a surface.

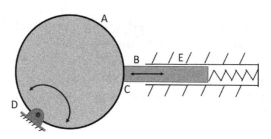

Figure 7.4: An example of the j_1 joint, similar to those used in internal combustion engines.

A has two degrees of freedom because the rotational and the translational movements are independent.

Note that we do not have to have the upper boundary of the groove. It can be an open surface if the rigid body is heavy enough to maintain contact with the surface, as shown in Fig. 7.3(b). The contact conditions in Fig. 7.3 are called sliding contact or contact with slip.

A rigid body assembly with a sliding contact is shown in Fig. 7.4. A circular rigid body (A) is rotating around a pin (D), while a rod (B) is pressed against the circular body at point C by a spring. The rotation of A will cause rod B to move horizontally, and the spring ensures the contact between A and B. Rod B can only move horizontally because it is constrained by the slot E. A practical example of Fig. 7.4 is the cam system commonly used in internal combustion engines to control the opening and closing of the intake and exhaust valves.

For the assembly of Fig. 7.4, we have three rigid bodies (A, B, D), two j_2 joints (D and E), and one j_1 joint (C). Applying Gruebler's equation, we have

$$M = 3(3) - 3(1) - 2(2) - 1(1) = 1 \tag{7.4}$$

Therefore, the assembly shown in Fig. 7.4 has only one degree of freedom.

7.1.2 Paradox of Gruebler's equation

Although Gruebler's equation provides a convenient way to determine the degrees of freedom for rigid body assemblies, the calculation results of Gruebler's equation are not always correct. As shown in Fig. 7.5, two examples are provided to demonstrate the flaw of Gruebler's equation. In Fig. 7.5(a), the assembly could be a set of frictional wheels or a pair of engaged gears.

Applying Gruebler's equation, the assembly of Fig. 7.5(a) should have zero degree of freedom because

$$M = 3(3) - 3(1) - 2(3) - 1(0) = 0 \qquad (7.5)$$

where the three j_2 joints are the two pin joints and the rolling contact between the gears. Similarly, for the mechanism in Fig. 7.5(b), Gruebler's equation predicts zero degrees of freedom, as

$$M = 3(5) - 3(1) - 2(6) - 1(0) = 0 \qquad (7.6)$$

However, both assemblies in Fig. 7.5 have one degree of freedom. This paradox is caused by a fundamental flaw of Gruebler's equation, which is that Gruebler's equation does not consider the relationship between the joints. Some of the joints might be redundant, such as the links in Fig. 7.5(b). If we remove one of the pined elements, there is no change in the motion of the assembly.

Another way to look at the paradox is that Gruebler's equation is a scalar equation, but the assembly is in 2D or 3D. As a result, similar joints in different orientations could have different effects in eliminating degrees of freedom. As shown in Fig. 7.6, rigid body D in assembly (a) has a partial rotational motion, while the rotation is completely eliminated

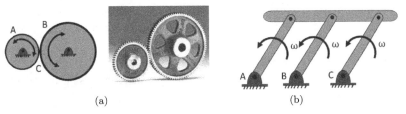

(a)   (b)

Figure 7.5: Rigid body assemblies which have more degrees of freedom than those predicted by Gruebler's equation: (a) a pair of friction wheels or gears and (b) a platform supported by multiple parallel links.

Source: (a) https://www.divilabs.com/2013/06/gears-their-common-types-used-in.html.

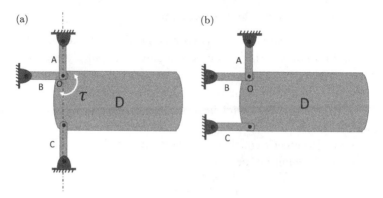

Figure 7.6: Examples to show the flaw of Gruebler's equation: (a) an incorrectly predicted case and (b) a correctly predicted case.

in (b); however, Gruebler's equation will predict zero degrees of freedom for both,

$$M = 3(5) - 3(1) - 2(6) - 1(0) = 0 \tag{7.7}$$

Note that there are two j_2s at point O for pinning three rigid bodies, A, B, and D, together.

However, the joints related to element C in assemblies of Figs. 7.6(a–b) have different constraining effects. As elements A, B, and C are all pined–pined rods, they only provide constraint forces along their element directions. In assembly of (a), all three constraint forces will pass through point O. These elements cannot resist a moment, τ, until they shift out of their current positions to make element C not point to O anymore. As a result, rigid body D can have a small partial rotational motion. On the other hand, the elements in assembly of (b) can resist any forces or moments applied to rigid body D to keep it from any movements. Because the arrangements of element C are different, they have different constraining effects. However, Gruebler's equation, as a scalar equation, cannot account for these effects.

In conclusion, we need to use Gruebler's equation with caution. With attention to potential paradoxes, Gruebler's equation is still an effective tool for structural designs.

7.1.3 Basic truss system design

We will now use Gruebler's equation to design basic structures of truss systems. As discussed above, truss systems are made of beams and pins.

Figure 7.7 illustrates how a stable truss system can be constructed. If we connect elements with pins to form a closed structure, as shown in Figs. 7.7(a–c), only the one with three elements forms a stable structure with zero degree of freedom as affirmed by Gruebler's equation. In the design of (a), the entire structure is rigid to maintain the triangular shape of the structure. On the other hand, in designs (b) and (c), the shapes of the structures can be changed, and thus these are not suitable for structures used for buildings or bridges.

The triangular basic truss design of Fig. 7.7(a) can be repeated to form more complicated structures, as shown in Figs. 7.7(d–e). We can simply add two beams at a time to form a new stable triangular structure, just like beams D and E are added to basic structure A-B-C in Fig. 7.7(d). Similarly, beams F and G can be added to structure A-B-C-D-E in Fig. 7.7(e). Note that even though we add two beams at a time to the existing stable structure, the newly added section does not have to be triangular. The new section formed by beams H and I in Fig. 7.7(e), along with beams G and E, is not a triangle, but it is stable.

From Fig. 7.7, we learn that the most basic truss system is the triangular arrangement. In this triangular arrangement, element A is fixed to the foundation as a stationary element (Fig. 7.8(a)). However, how to fix element A is not a trivial question. One way to do this is to apply pins to both ends of element D as shown in Fig. 7.8(b). If we do so,

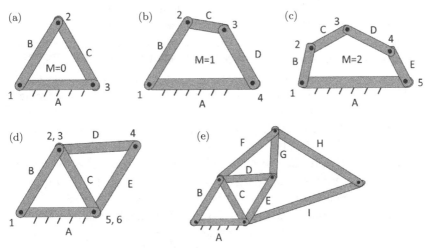

Figure 7.7: Truss system designs with different numbers of elements: (a) a stable triangular truss system; (b) an unstable four-member truss system; (c) an unstable five-member truss system; (d–e) a stable truss system with triangular subsystems.

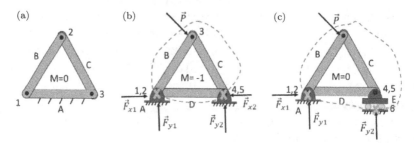

Figure 7.8: (a) A triangular truss system with one fixed member; (b) a triangular truss system with two fixed pins; (c) a triangular truss system with one fixed pin and one floating pin.

there are some practical concerns. First, Gruebler's equation will show that the truss system has a negative degree of freedom, indicating that the system is now overconstrained. Furthermore, when both ends of element D are fixed by pins, element D is not allowed to expand thermally in the longitudinal direction. With a large structure, the thermal expansion could be substantial, causing element D to deform.

In general, we would like the structure to be deterministic so that we can calculate the loading of each member. For the structure of Fig. 7.8(b), when the truss system is subjected to an external force, $\vec{P}$, there are two reaction forces at each pin, for four unknowns. Because we have only three force equilibrium equations, we cannot determine the value of these four unknown forces. On the other hand, if we add a rolling pin (E) to the truss system, as shown in Fig. 7.8(c), element D is no longer restricted. Not only can it accommodate thermal expansion in the longitudinal direction, we also have only three unknown reaction forces, which can be determined deterministically. Using Gruebler's equation, we have five elements (A, B, C, D, and E), and six j_2 joints, which zero degrees of freedom, forming a properly constrained truss system.

The basic 2D truss design can be extended to form 3D structures. A common practice is to place the same truss systems in parallel and then connect them rigidly (not through truss element), as is often seen in the construction of bridges and buildings. We can also create 3D truss systems with truss elements connected together by ball joints. The basic 3D truss system will be a cone structure, as shown in Fig. 7.9(a). Similar to the 2D truss system, the basic 3D truss cone can be extended, now by adding three members at a time, as shown in Fig. 7.9(b). The 3D truss cone systems are apparently not widely used due to the complexity, cost, and difficulty in space management.

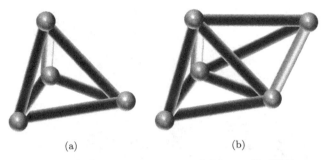

(a) (b)

Figure 7.9: (a) A basic 3D cone truss system and (b) a stable 3D truss system with cone subsystems.

We thus only focus on the 2D truss system for analysis in the following sections.

7.1.4 Force analysis for truss systems

Truss systems are widely used for bridges, roofs, and hoist structures. A historical review of truss system design can be found on the Internet.[1] We will present a method to conduct truss loading analysis for 2D truss systems, using the free-body diagram analysis method presented in Chapters 5 and 6.

For 2D truss systems, we only have three equilibrium equations: two force equations and one moment equation. Therefore, when we draw a boundary to isolate a free-body, the free-body diagram cannot contain more than three unknown constraint forces. If the free-body diagram contains only one pin, all the constraint forces will intersect at the center of the pin. As a result, we only have two equilibrium force equations; therefore, we can only have two unknowns.

To analyze the entire system, we first draw a boundary around the entire truss system to determine the reaction forces of the supports at the foundation. From there, we can draw boundaries, one pin at a time and, gradually, we can determine the force of each truss element. We will describe the method via the example of a Warren truss bridge, shown in Fig. 7.10. The Warren truss system uses equilateral triangles,[2] which simplify the inventory of beams.

As shown in Fig. 7.10, we first draw a boundary for the entire system to isolate it from the foundation. There is a pin constraint at A and a point contact at F.

---

[1]https://www.garrettsbridges.com/design/trussdesign/.
[2]https://www.garrettsbridges.com/design/warren-truss/.

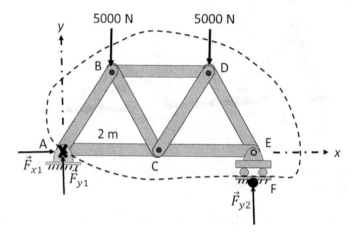

Figure 7.10:   A simple Warren bridge truss system for force analysis.

Applying the ABCC method, based on the FBD reference table in Chapter 6,

(A) — We have two applied forces, 5000 N each at points B and D, respectively;

(B) — We neglect the body force;

(C) — There is a normal contact force at point F;

(D) — There are two constraint forces at point A.

Now applying the equilibrium equations, with the moment equation with respect to point A, we have,

$$\sum F_x = 0 : F_{x1} = 0 \tag{7.8}$$

$$\sum F_y = 0 : -5000 - 5000 + F_{y1} + F_{y2} = 0 \tag{7.9}$$

$$\sum (\vec{\tau})_A = 0 : -5000 \times 1 - 5000 \times 3 + F_{y2} \times 4 = 0 \tag{7.10}$$

Solving the above equations, we obtain, $F_{x1} = 0$, $F_{y1} = 5000\,\text{N}$, and $F_{y2} = 5000\,\text{N}$.

Now, we can isolate each pin to solve for the force of each element. However, because we only have two equilibrium force equations, the pin we isolate can only be involved with two unknown truss members. At a quick glance, we notice that pin A has two truss members, B, three members, C, four members, D, three members, and E, two members. Therefore, we can start at pin A to solve for the forces for members AB and AC. After that, pin B will be involved with only two unknown members (BC and BD).

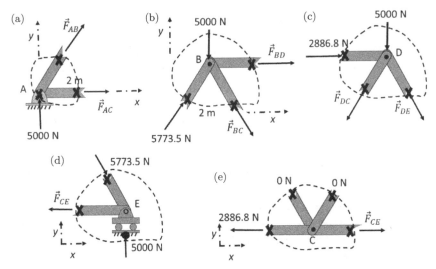

Figure 7.11: Free body diagram analysis for determining all the element forces for the Warren bridge of Fig. 7.10: (a–e) FBDs for joints A, B, D, E, and C, respectively.

After that, pins C and pin D will be involved with two unknown members. The entire procedure is shown in Fig. 7.11.

For pin A, we draw new boundary as shown in Fig. 7.11(a). We will go through the same ABCC method and mark two constraint forces along elements AB and AC, respectively. We can just assume these constraint forces as tensile; thus, the forces are pointing away from the boundary, as shown in Fig. 7.11(a).

$$\sum F_x = 0 : F_{AB} \times \cos(60°) + F_{AC} = 0 \tag{7.11}$$

$$\sum F_y = 0 : 5000 + F_{AB} \times \sin(60°) = 0 \tag{7.12}$$

Solving the above equations, we obtain $F_{AB} = -5773.5\,\text{N}$ and $F_{AC} = 2886.8\,\text{N}$. The negative value of F_{AB} indicates that element AB is actually in compression, different from the assumption we made initially, while the positive value of F_{AC} indicates that it is indeed in tension.

Once we have determined the forces of elements AB and AC, there are only two unknowns related to pin B. Therefore, we can proceed to draw a free-body diagram around pin B, as shown in Fig. 7.11(b). Solving the corresponding equilibrium equations, we obtain $F_{BC} = 0\,\text{N}$ and $F_{BD} = -2886.8\,\text{N}$. It is interesting to note that the force on element BC is zero, while element BD is in compression.

We can then consider pin D with the information of element BD. Now, there are only two unknowns, elements DC and DE. Solving the equilibrium equations, we obtain $F_{DC} = 0\,\text{N}$ and $F_{DE} = -5773.5\,\text{N}$.

We then proceed to pin E, now with only one unknown for element CE. We obtain $F_{CE} = 2886.8\,\text{N}$.

Finally, if we still consider element CE as unknown, we can draw a boundary for pin C and calculate the force for element CD again. We obtain the same result as before. This verifies the overall calculation.

Typically, we need to pay attention to the element that is subjected to the largest force. Due to the concern of buckling, we should also verify the elements under compression to assess if buckling failures would occur. Buckling is a geometrical unstable phenomenon when a long and slender element is subjected to compressive loading. The element could fail at loads well below its material strength. The buckling calculation is beyond the scope of this book.

It is again interesting to note that elements BC and DC are not subjected to forces in this loading condition. However, it does not mean that these two elements can be eliminated. If we add a horizontal force to pin D, and redo the entire calculation, the force on element BC is still zero, but the force on element DC is now 5000 N of tensile force.

The loadings of most truss systems can be determined via the above method by going through every pin of the system. However, this method could be time-consuming. It is also possible to determine the loadings to some internal elements by choosing a free-body diagram wisely. The example below illustrates this method. As shown in Fig. 7.12, a K truss bridge is shown. As discussed earlier, long and slender elements are susceptible to buckling failure when they are subjected to compressive

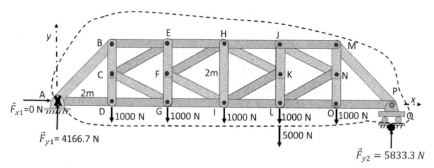

Figure 7.12: A K truss bridge example, subjected to distributed loading and a concentrated loading.

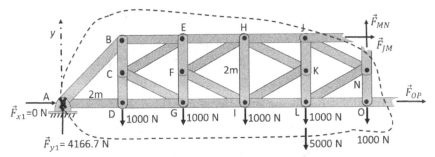

Figure 7.13: A free-body diagram with its boundary cutting through members JM, MN, and OP.

loadings. The K truss system shortens some elements to reduce the likelihood of bucking failures.

The bridge in Fig. 7.12 is subjected to distributed loading, represented by the 1000 N force acting at pins, D, G, I, L, and O. In addition, a concentrated force of 5,000 N is acting at pin L. The constraint forces at the supports can be easily determined and are treated, as shown in Fig. 7.12. If we try to isolate a sub-system, the boundary cannot cut through more than three members, and these three members cannot point to one point. A suitable boundary is shown in Fig. 7.13, which cuts through members JM, MN, and OP, so that the corresponding three unknown member forces, F_{MN}, F_{JM}, and F_{OP}, do not intersect at a single point.

Now, write down the three equilibrium equations and solve for them,

$$\sum F_x = 0: \ F_{OP} + F_{JM} = 0 \tag{7.13}$$

$$\sum F_y = 0: \ 4166.7 - 10000 + F_{MN} = 0 \tag{7.14}$$

$$\sum (\vec{\tau})_A = 0: \ -1000 \times 2 - 1000 \times 4 - 1000 \times 6 - 6000 \times 8$$
$$- 1000 \times 10 + F_{MN} \times 10 - F_{JM} \times 2 = 0 \tag{7.15}$$

We obtain $F_{MN} = 5833.3\,\text{N}$, $F_{OP} = 5833.3\,\text{N}$, and $F_{JM} = -5833.3\,\text{N}$.

In some special cases, we can draw a boundary to cut through more than three members. In such a case, we might be able to determine some of the forces. Figure 7.14 illustrates this method.

As shown in Fig. 7.14, there are four unknown forces of members EH, EF, FG, and GJ. Because we only have three equilibrium equations, we cannot solve for all these four unknown forces. However, we notice that three of these unknown forces intersect at point G. Therefore, if we write down

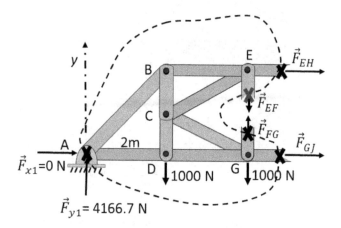

Figure 7.14: A free-body diagram with its boundary cutting through four members.

an equilibrium moment equation with respect to point G, this equation will contain only one unknown, F_{EH}, which can be solved.

$$\sum (\vec{\tau})_G = 0: -4166.7 \times 4 + 1000 \times 2 - F_{EH} \times 2 = 0 \qquad (7.16)$$

We obtain, $F_{EH} = -7354\,\text{N}$, which is in compression. We can then quickly determine $F_{GJ} = 7354\,\text{N}$ with the equilibrium force equation in the x direction. However, we are not able to determine the values of F_{EF} and F_{FG} with this free-body diagram. The pin-by-pin method can be used if we are interested in determining these two forces.

In conclusion, with effort, it is almost always possible to determine all the forces in a truss system.

7.1.5 Force analysis for mixed truss systems

There are some mechanical systems that utilize many truss members but also some non-truss members. It is important that one, after learning the previous sections, does not think all elements in all machines are truss elements. It is very important to observe that a truss member, with the pined–pined constraint, must not have any forces acting on the member except at the two pins. Two applications of truss members with mixed members are shown in Fig. 7.15. In Fig. 7.15(a), a mechanical scissor jack has seven members, A–G. Members A–E are truss members, joined by pins only, with forces along the member longitudinal direction. However, member F and G are not truss members. For example, the forces acting on G will be

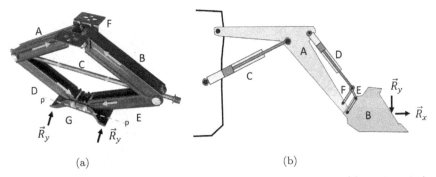

(a)   (b)

Figure 7.15: Two examples of truss systems with mixed members: (a) a scissor jack and (b) the front loader of a tractor.

Source: (a) http://images3.campingworld.com/CampingWorld/images/products/5000/originals/55637n.jpg.

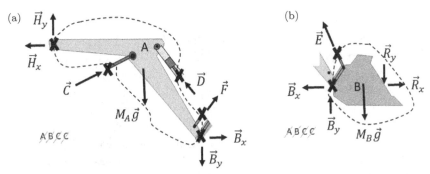

Figure 7.16: Free-body diagram analysis using the ABCC method for (a) member A and (b) member B of the tractor in Fig. 7.15.

mostly vertical, not horizontal along the direction of the two pins (shown as a dash line p–p). There are other external forces acting on member G, such as at the points where both feet touch the ground, not limited to the pins, as required by the pined–pined rod constraint.

For the front loader of a tractor in Fig. 7.15(b), six members, A–F, are identified. Among them, only members C, D, E, and F are truss members, while members A and B are not. A detailed free-body diagram analysis for the tractor of Fig. 7.15(b) is shown in Fig. 7.16.

To construct the free-body diagram of member B of Fig. 7.15(b), we draw a boundary to isolate it as shown in Fig. 7.16(b). We then go around the boundary to mark the constraints (no contacts in this case), which include a pin constraint between A and B and a pined–pined rod constraint

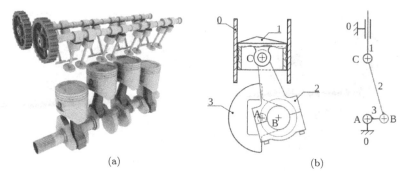

(a)   (b)

Figure 7.17: (a) The piston assembly of an internal combustion engine and (b) the
mechanism of the piston system represented with mixed truss members.
Source: (a) https://upload.wikimedia.org/wikipedia/commons/6/69/Engine_movingpa
rts.jpg; (b) https://upload.wikimedia.org/wikipedia/commons/thumb/f/f7/Piston_bie
lle_vilebrequin_coupe_et_schema_cinematique.svg/1280px-Piston_bielle_vilebrequin_coup
e_et_schema_cinematique.svg.png.

of E. Now, write down ABCC to account for all forces. There are two applied
forces, $\vec{R}_x$ and $\vec{R}_y$, (crossing out letter A), one body force, $M_B\vec{g}$, at the mass
center of member B, (crossing out letter B), no contact forces (crossing out
letter C), and two pin constraint forces, $\vec{B}_x$ and $\vec{B}_y$, and one pined–pined
rod constraint force, $\vec{E}$, based on Fig. 7.16(b). If we know $\vec{R}_x$, $\vec{R}_y$, and $M_B\vec{g}$,
we can solve for three unknown forces, $\vec{B}_x$, $\vec{B}_y$, and $\vec{E}$. We can construct
the free-body diagram for member A accordingly, as shown in Fig. 7.16(a).
With the forces $\vec{B}_x$ and $\vec{B}_y$ obtained from free-body diagram Fig. 7.16(b),
if we can determine the forces of $\vec{C}$ and $\vec{D}$ based on the hydraulic pressure,
and the weight of member A, $M_A\vec{g}$ we can solve for three unknowns of $\vec{H}_x$,
$\vec{H}_y$, and $\vec{F}$. Note that the hydraulic cylinders of C and D are truss members
because they are only pined at both ends and have no other forces acting
on them. The hydraulic forces are aligned with the longitudinal direction
of the cylinder.

Sometimes, it is not obvious if a member is a truss member. As shown
in Fig. 7.17, the piston assembly of an internal combustion engine can be
illustrated as a linkage system, as shown in Fig. 7.17(b). In this equivalent
linkage system, we have linkage AB (crankshaft) and linkage BC (piston
rod). The piston rod (BC) can be considered as a truss member because it
is only pined at both ends. However, the linkage AB (crankshaft) is not a
pined–pined rod because pin A is not a pin free to rotate. The "pin" A is
the shaft of the crank, which is connected to the axle to provide a torque.
Therefore, this "pin" is not friction-free as required in a pined–pined rod.

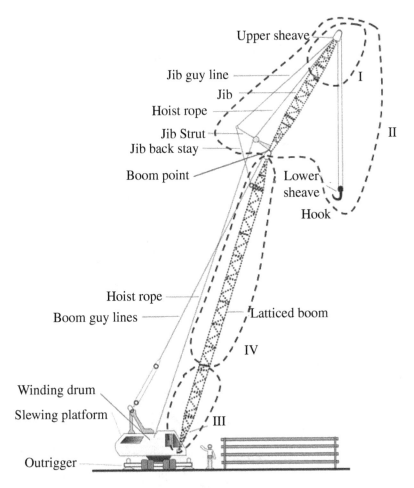

Figure 7.18: The boundaries defined for analyzing a crane system.
Source: https://upload.wikimedia.org/wikipedia/commons/thumb/4/4a/Crane_machin e_slewing_platform.svg/1200px-Crane_machine_slewing_platform.svg.png.

The force on the crankshaft linkage AB will not be along line AB. The force on the piston rod, on the other hand, can be assumed to be along line BC.

Another interesting example of a mixed truss system is a crane machine used in construction, as shown in Fig. 7.18. Such a crane system has a latticed boom, held by a boom guy line, a jib, held by a jib guy line, and a hoist rope with a hook. The boom and the jib are long and slender, held by pins.

Can the boom and the jib be considered as truss members with forces only acting along the longitudinal directions without a bending moment?

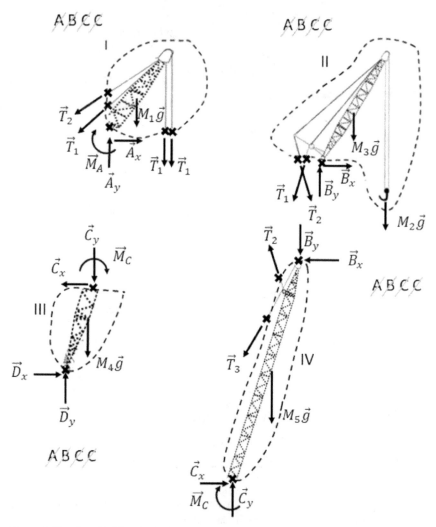

Figure 7.19: Detailed free-body diagram analysis using the ABCC method for the boom and the jib of the crane system of Fig. 7.18.

To answer this question, we first take a conservative approach by not considering them as truss members. We draw four free-body diagrams for the jib and the boom, as shown in Fig. 7.19. In these detailed free-body diagrams, we do not assume the jib and the boom as truss members. As a result, as shown in free-body diagrams I, III, and IV, there are bending moments at the cross-sections of the jib and the boom when their weights are considered. There are many forces and moments involved in these free-body diagrams.

For free-body diagram I, we can determine the tension of hoist rope $(\vec{T}_1)$ based on the weight on the hook $(M_2\vec{g})$ (free-body diagram II). We can also estimate the weight of the partial jib in free-body diagram I $(M_1\vec{g})$. There are still four unknowns, $\vec{T}_2$ (jib guy line tension), $\vec{A}_x$, $\vec{A}_y$, and, $\vec{M}_A$.

However, for free-body diagram II, if we know $M_2\vec{g}$ and $M_3\vec{g}$ (the weight of the jib), we can actually solve for the three unknowns, $\vec{T}_2$, $\vec{B}_x$, and $\vec{B}_y$. From here, we can go back to free-body diagram I and solve for $\vec{A}_x$, $\vec{A}_y$, and $\vec{M}_A$.

For free-body diagram III, if we know the partial boom weight, $M_4\vec{g}$, we still have five unknowns, $\vec{C}_x$, $\vec{C}_y$, $\vec{D}_x$, $\vec{D}_y$, and $\vec{M}_B$ — too many to solve.

For free-body diagram IV, if we know the partial boom weight, $M_5\vec{g}$, we still have four unknowns, $\vec{T}_3$ (boom guy line tension), $\vec{C}_x$, $\vec{C}_y$, and $\vec{M}_C$ — again, too many to solve. If we can measure the boom guy line tension, $\vec{T}_3$, it is possible to solve for $\vec{C}_x$, $\vec{C}_y$, and $\vec{M}_C$. We can then go back to free-body diagram III to solve for $\vec{D}_x$ and $\vec{D}_y$.

If we choose to use lightweight materials for the jib and the boom, we can neglect their weight in our analysis, and the free-body diagram analysis becomes much simpler. However, bear in mind that we would underestimate the loadings.

By assuming both the jib and the boom as under the pined–pined rod constraints, the free-body diagrams are greatly simplified, as shown in Fig. 7.20.

For free-body diagram I of Fig. 7.20, we can determine the tension of the hoist rope $(\vec{T}_1)$ based on the weight on the hook $(M_2\,\vec{g})$ (free-body diagram II). There are only two unknowns, $\vec{T}_2$ (jib guy line tension) and $\vec{A}$, readily solvable.

Now, for free-body diagram II, we can solve for the two unknowns $\vec{B}_x$ and $\vec{B}_y$.

For free-body diagram IV, we have only two unknowns, $\vec{T}_3$ (boom guy line tension) and $\vec{C}$, readily solvable. Finally, we can then go back to free-body diagram III, to solve for $\vec{D}_x$ and $\vec{D}_y$.

If we are designing such a crane system, the first approach is to use the simplified analysis of Fig. 7.20. Then, the weight of the jib and the boom can be added to determine the bending moments as in Fig. 7.19. The results are then compared with those from Fig. 7.20. This allows us to

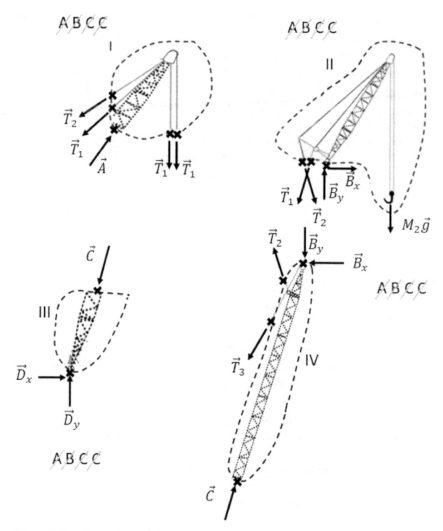

Figure 7.20:   Approximated free-body diagram analysis using the ABCC method for the boom and the jib of a crane system.

determine the acceptable weight of the boom and the jib for keeping the bending moments low.

Finally, it should be pointed out that the outrigger of the crane machine must be able to provide sufficient force so that the entire crane machine will not flip over.

7.2 Hand Tools

Many kinds of hand tools have been invented to assist in assembling, installing, and repairing jobs. In Figs. 7.21–7.23, many common tools, including wrenches, pliers, screwdrivers, etc., are shown. Mechanics will tell you that they are always one tool short. Many special tools are also developed for difficult jobs due to, for example, limited space.

From the Statics point of view, these hand tools are used to transmit forces/torques, to convert between forces and torques, and to amplify forces/torques. We will choose a few common tools and conduct force analysis for them.

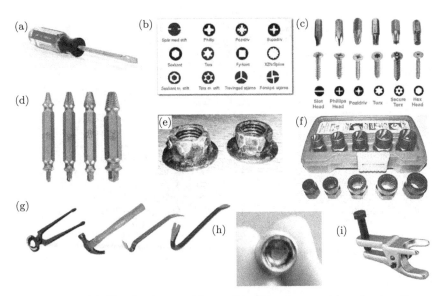

Figure 7.21: (a) Slot head screwdrivers; (b) different screwdriver head patterns; (c) screwdriver bits and screws; (d) screw extractors; (e) rounded nuts; (f) bolt extractors; (g) pry bars; (h) rounded allen bolts; (i) ball joint separators.

Source: (a) https://upload.wikimedia.org/wikipedia/commons/thumb/1/1c/Screw_Dri ver_display.jpg/1200px-Screw_Driver_display.jpg; (b) https://lh3.googleusercontent.com /proxy/7oZZ62Ywb1PxdvFt-E2af-9EcCjWhGf_K5_hhNq5TvV9zTahSdsq_4cQvSYznfl4 1gJiixkZkmx7edxNmlZOSX0HHX2QA-pqDAY72MgY=s0-d; (c) http://wiki.dtonline. org/images/thumb/8/8c/ScrewHeadTypes.png/300px-ScrewHeadTypes.png; (d) https: //i.stack.imgur.com/rryDy.jpg; (e) https://i.stack.imgur.com/B86u8.png; (f) http:// www.crankshaftcoalition.com/wiki/images/thumb/2/2e/Sockets_003.jpg/300px-Sockets _003.jpg; (g) http://wiki.dtonline.org/images/thumb/0/0c/PincersCrowBarClaw.png/ 800px-PincersCrowBarClaw.png; (h) https://i.stack.imgur.com/14SAs.jpg; (i) https:// i.stack.imgur.com/n32Xs.jpg.

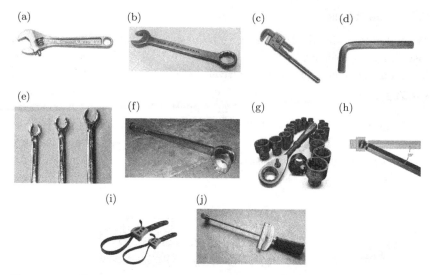

Figure 7.22: Common wrenches: (a) adjustable wrenches; (b) open-end and box-end combination wrenches; (c) pipe wrenches; (d) allen keys; (e) flare wrenches; (f) ratchet wrench with sockets (12 points); (g) spline socket set and ratchet wrench; (h) ball end allen keys; (i) rubber strap wrenches; (j) torque wrenches, beam type.

Source: (a) https://i.stack.imgur.com/bPMfo.jpg; (b) https://upload.wikimedia.org/wikipedia/commons/thumb/4/4c/1933_Plomb_Combination_Wrench.jpg/1200px-1933_Plomb_Combination_Wrench.jpg; (c) https://i.stack.imgur.com/Uai0Y.jpg; (d) https://upload.wikimedia.org/wikipedia/commons/5/5e/Kluc_imbus.jpg; (e) http://i.stack.imgur.com/ZTKtS.jpg; (f) https://upload.wikimedia.org/wikipedia/commons/thumb/9/9a/Click-torque-wrench.jpg/1200px-Click-torque-wrench.jpg; (g) http://i.stack.imgur.com/2YWNj.png; (h) https://i.stack.imgur.com/kgcQW.jpg; (i) https://i.stack.imgur.com/em5gR.jpg; (j) https://upload.wikimedia.org/wikipedia/commons/c/cc/Western_Forge_Craftsman_beam_torque_wrench.jpg.

7.2.1 Force-Wrench (screwdrivers)

The force applied to a screwdriver is a good example of a force and a wrench described in Fig. 2.11, Chapter 2. When using a screwdriver, we press the screwdriver against the head of a screw and twist to tighten or remove a screw. Often, a cordless driver is used with a driver bit to speed up the work and to apply higher torques. The most common screw head patterns are the slot head and Phillips. The slot head typically offers a higher torque capacity, but it is harder to align the screwdriver with the slot. The Philips screws, on the other hand, are easy to align, but we have all had bad experience of stripping Philips screws. Many different patterns are then developed to improve the alignment and the torque capacity, such

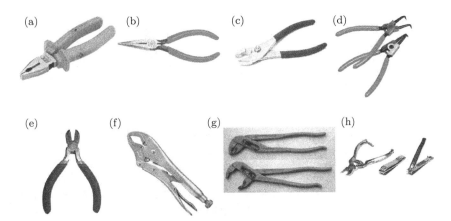

Figure 7.23: Common pliers: (a) lineman pliers; (b) long nose pliers; (c) slip joint pliers; (d) ring spreader pliers; (e) diagonal plier; (f) locked plier; (g) tongue and groove pliers; (h) clipper pliers and nail clippers.

Source: (a) http://www.pngall.com/wp-content/uploads/2016/05/Plier-PNG-Image.p ng; (b) https://upload.wikimedia.org/wikipedia/commons/thumb/8/8e/Klein_Tools_D 203-6.jpg/1200px-Klein_Tools_D203-6.jpg; (c) https://upload.wikimedia.org/wikipedia/ commons/thumb/5/51/Tool-pliers.jpg/1200px-Tool-pliers.jpg; (d) https://i.stack.imgur .com/ThrX6.jpg; (e) https://upload.wikimedia.org/wikipedia/commons/f/fd/Cutting_ tool_1a.jpg; (f) https://i.stack.imgur.com/L0MMO.png; (g) https://upload.wikimedia. org/wikipedia/commons/thumb/2/29/WaPuZa_Rothenberger_03.jpg/1200px-WaPuZa_ Rothenberger_03.jpg; (h) https://upload.wikimedia.org/wikipedia/commons/thumb/4/ 4d/Nail-clippers-variety.jpg/300px-Nail-clippers-variety.jpg.

as hex, torx, or XZN type, as shown in Figs. 7.21(b–c). When a screw is stripped, there are extractors used to remove it (Fig. 7.21(d)).

7.2.2 Force amplification

There are many ways to amplify a force. In this section, we will present several examples of force amplification via the use of leverages, wedges, pulleys, and hydraulics.

Leverage: When we use a plier to grip or to cut an object. The force needed to achieve the gripping or cutting is typically higher than the strength of our hands. The plier offers a force amplification (Fig. 7.23). We can conduct a force analysis to explain this force amplification, as shown in Fig. 7.24. In Fig. 7.24(a), we draw a boundary for the entire plier and the piece to be cut. Going through the ABCC method, we only have two applied forces, $\vec{C}$, on the handle applied by hands. These two forces must be equal in size and in the opposite directions so that an equilibrium is achieved. Therefore,

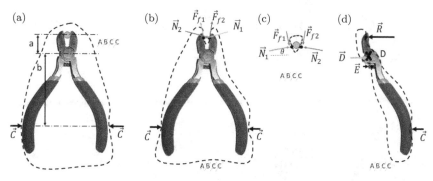

Figure 7.24:   (a) FBD for the plier with an object; (b) FBD for the plier only; (c) FBD for the object only; (d) FBD for one-half of the plier.

they must be horizontal as shown in Fig. 7.24(a). Now, if we draw a new boundary to exclude the piece, as shown in Fig. 7.24(b), we have two contact points. At one contact point, we have the normal contact force, $\vec{N}_1$ and the frictional force, $\vec{F}_{f1}$, and similarly, $\vec{N}_2$, and $\vec{F}_{f2}$, for the other contact point. These forces must cancel each other. More specifically, the resultant force of $\vec{N}_1$ and $\vec{F}_{f1}$, denoted as $\vec{R}$, should be equal to the resultant force of $\vec{N}_2$ and $\vec{F}_{f2}$. These resultant forces will again be horizontal and in opposite directions. Now, if we draw a boundary for the piece to be cut, as shown in Fig. 7.24(c), there will only be two contact points, with the same normal contact forces and frictional forces. For this plier, because there is a contact angle, θ, the normal contact forces, $\vec{N}_1$ and $\vec{N}_2$, are not horizontal. We need the frictional forces, $\vec{F}_{f1}$ and $\vec{F}_{f2}$, to make the resultant forces, $\vec{R}$, horizontal. However, there is a limit to this frictional force, which may not be sufficient to keep the piece in place. We all have had this experience, as shown in Fig. 7.24, of a large round piece slipping out when we squeeze the plier handles to cut it. Finally, we construct a free-body diagram by drawing a boundary around just one side of the plier, as shown in Fig. 7.24(d). We have two contact points, one at the piece to be cut and one at the spring of the plier, and one pin constraint. The applied force, $\vec{C}$, and the contact forces, $\vec{R}$ and $\vec{E}$, are all horizontal. The constraint force, $\vec{D}$, at the pin should be horizontal too. Typically, the contact force from the spring, $\vec{E}$, is small compared with the gripping force, $\vec{C}$, and the cutting force $\vec{R}$, and it can be ignored. Now, the moment equilibrium equation with respect to point D is

$$R = \left(\frac{b}{a}\right) C \tag{7.17}$$

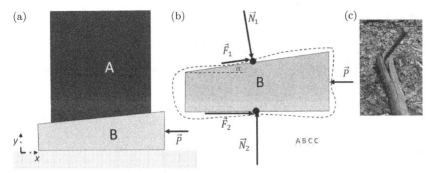

Figure 7.25: (a) A wedge to life a heavy object; (b) FBD for the wedge; (c) ax as a wedge to split wood.
Source: (c) https://upload.wikimedia.org/wikipedia/commons/5/51/Merlin_usteye_bok ion.jpg.

The force amplification factor is $\frac{b}{a}$. To increase the amplification, we can move the piece closer to point D to reduce the length of a, or to grip the plier further out to increase the length of b. This matches the common experience we have. However, this force amplification is not free. The force at the pin, $\vec{D}$, is equal to the sum of $\vec{C}$ and $\vec{R}$. This pin must be strong enough to withstand this sum force.

Wedges: Another way of force amplification is through shims and wedges. As shown in Fig. 7.25(a), a heavy object A is to be lifted by a wedge B with the application of force $\vec{P}$.

A free-body diagram analysis for wedge B is shown in Fig. 7.25(b). Deriving the equilibrium force equation in the horizontal direction, we get

$$\sum F_x = 0 \colon N_1 \sin(\alpha) + F_1 \cos(\alpha) - P + F_2 = 0 \qquad (7.18)$$

We also have the friction force and the normal force relationships as,

$$F_1 = \mu N_1 \qquad (7.19)$$
$$F_2 = \mu N_2 \qquad (7.20)$$

It is also easy to see that N_2 is the total weight of objects A and B, denoted as W. Now, we have

$$P = N_1(\sin(\alpha) + \mu \cos(\alpha)) + \mu W \qquad (7.21)$$

The action of wedging is more effective if the friction is low, which can be achieved by lubrication. If the friction is negligible, the above equation can be simplified to

$$P = N_1 \sin(\alpha) \tag{7.22}$$

The actual lifting force, N_L, against object A is the vertical component of N_1, therefore,

$$N_L = P\, \frac{\cos(\alpha)}{\sin(\alpha)} = \beta P \tag{7.23}$$

where $\beta = \frac{\cos(\alpha)}{\sin(\alpha)}$ is the force amplification factor.

The following table shows the amplification factor for different values of α.

α (degree)	β
2.5	22.9
5	11.4
10	5.7
15	3.7

The cost of force amplification is the reduction in the lifting movement. The lifting movement of object A is $1/\beta$ of the horizontal movement of wedge B. This wedge action can be applied to split wood with an axe, as shown in Fig. 7.25(c).

Pulleys: Cables and pulleys are also used for force amplification. In a pulley and cable assembly, a few practical assumptions are made. First, the cable or wire is assumed to be non-stretchable. A cable is flexible in that it bends, but when being stretched, its length stays the same. Second, the pulley can rotate freely, i.e. the friction is negligible, like a pin constraint. The weight of the pulley is also considered negligible.

With these assumptions, if we draw a free-body diagram of a pulley and the wire around it, as shown in Fig. 7.26(a), we can derive a moment equilibrium equation with respect to the center of the pulley as

$$-T_1 r + T_2 r = 0 \tag{7.24}$$

where r is the radius of the pulley. As a result, $T_1 = T_2$. This result indicates that with the same wire running through any number of pulleys, the tension of this wire at any point stays the same. This is pretty much what we need to describe how a pulley/cable system amplifies a force.

We can apply this result to a bicycle hanging device shown in Fig. 7.26(b). There are four pulleys, A, B, C, and D. Pulleys B and C are

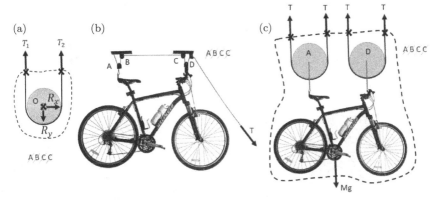

Figure 7.26: (a) FBD for a pulley; (b) a pulley system for bicycle storage; (c) FBD for the bicycle pulley system.

Source: (b) https://i.stack.imgur.com/oYHCL.jpg.

attached to the ceiling, while pulleys A and D are hooked to the bicycle. The bicycle is pulled up by pulling the cable with a force T, which is also the tension of the cable. The force amplification is illustrated by the free-body diagram shown in Fig. 7.26(c). The boundary is drawn to include the bicycle and pulleys A and D, cutting through the cable. It becomes clear that the pulling force T at the end of the cable becomes four times to lift the bicycle. If three floating pulleys together with three fixed pulleys are used to hook the bicycle, then the force amplification will be six times.

Hydraulics: Similar to the case wherein the tension is the same throughout a cable, the pressure of the hydraulic fluid is the same throughout the same hydraulic path. A simple example of a hydraulic jack and its exposed view are shown in Fig. 7.27.

Constructing a free-body diagram for each piston of the hydraulic jack, as shown in Fig. 7.27(b), we have,

$$A_1 p - F_1 = 0 \tag{7.25}$$

$$A_2 p - F_2 = 0 \tag{7.26}$$

The relationship between F_1 and F_2 becomes

$$\frac{F_2}{F_1} = \frac{A_2}{A_1} \tag{7.27}$$

The force amplification factor is the ratio of the piston areas. If the area of piston #2 is four times that of piston #1, then the amplification factor is

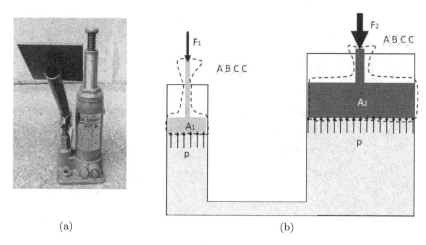

(a) (b)

Figure 7.27: (a) A hydraulic jack and (b) FBD for the hydraulic jack.

Source: (a) http://tech.txdi.org/t_tech/image/5000hydraulics/5000hy5.gif.

$\frac{A_2}{A_1} = 4$. The fluid should be at static equilibrium so that the same hydraulic pressure is maintained throughout the hydraulic path. If there is a high fluid flow, there will be pressure loss along the path, reducing the effective force amplification.

7.2.3 Force-to-torque conversion (wrenches)

When we use a wrench (see Fig. 7.22) to tighten or release a bolt, the force condition is more complicated, involving force/torque transmission, conversion, and amplification. We will describe these actions in detail.

The force analysis of a ratchet torque wrench is shown in Fig. 7.28(a). The force, $\vec{D}$, is applied at the handle. This force is converted to a torque or a moment, $\vec{M}$, through the leverage distance, a.

The ratchet gear mechanism is shown in Fig. 7.28(b). The gear of the ratchet set is also the drive for the wrench where a socket is attached. CCW rotations of the gear are not allowed through the action of a pawl. The force analysis of the ratchet gear is shown in Fig. 7.28(c). Going through the ABCC method, we have (A): applied forces and moments, $\vec{D}$ and $\vec{M}$; (B): no body forces by neglecting gravity; (C): contact forces between the gear and the pawl, $\vec{P}$; and (C): constraint forces at the pin support of the gear, $\vec{R}_x$ and $\vec{R}_y$. The gear is held in position through a short shaft, which is equivalent to a pin constraint.

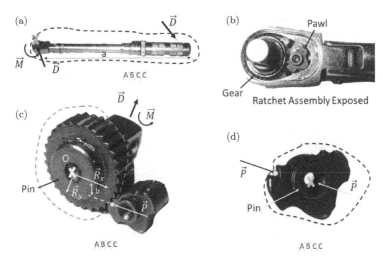

Figure 7.28: Force analysis of a ratchet wrench on force-torque conversion: (a) FBD for the entire torque wrench; (b) Internal gear assembly of the torque wrench; (c) FBD for the main gear; (d) FBD for the pawl.

Similarly, we can construct a free-body diagram for the pawl of the ratchet set as shown in Fig. 7.28(d). There is only one contact point and one pin constraint involved in the FBD of the pawl. We can neglect the little force of the spring and the ball behind the pawl. From Fig. 7.28(d), we can conclude that the contact force, $\vec{P}$, must pass through the center of the pin support and the constraint force at the pin must be equal and in opposite direction to $\vec{P}$. Now, looking at the free-body diagram of Fig. 7.28(c), we can readily derive the equilibrium equation of moment with respect to point O as

$$M = aD = bP \tag{7.28}$$

and

$$P = \frac{a}{b}D \tag{7.29}$$

Equation (7.29) indicates that the force $\vec{P}$ is amplified from force $\vec{D}$. The amplification can be in the range of 30 to 50. This large force $\vec{P}$ has to be sustained by the strength of the gear teeth. As a result, people often do not understand that a ratchet wrench should not be used to release a tight bolt because the force acting on the ratchet gear can exceed the strength of the gear teeth and damage it. A breaker bar is more suitable for releasing a tight or stuck bolt.

7.2.4 Torque-to-force conversion

The torque applied by a socket to a hex bolt is converted to forces at the bolt head, as shown in Fig. 7.29. Three pairs of force couples are produced, $(\vec{W}_1, \vec{W}_4)$, $(\vec{W}_2, \vec{W}_5)$, and $(\vec{W}_3, \vec{W}_6)$, which combine to produce the torque,

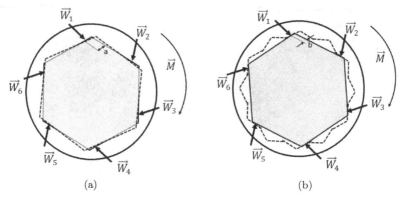

(a) (b)

Figure 7.29: Torque to force conversion when a socket applies a torque to a bolt: (a) for a 6-point socket and (b) for a 12-point socket.

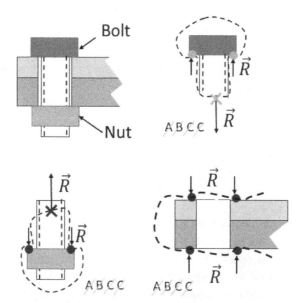

Figure 7.30: Torque-to-force conversion when a bolt and a nut are used to clamp two plates together.

$\vec{M}$. These force pairs are applied at different locations of the bolt head depending on the type of the socket used. In Fig. 7.29(a), a six-point socket is used, while in Fig. 7.29(b), a 12-point socket is used. Note that in a 12-point socket, the points of application of these forces are closer to the corners of the hex head. As a result, there is a greater risk of rounding a bolt head using a 12-point socket. Images of rounded bolts are shown in Fig. 7.21(e). This is the price paid for the easy fitting feature of 12-point sockets. There are many alternative designs to improve the 12-point socket design without sacrificing too much in loading capacity, while maintaining its easy fitting feature.

An important application using torque-to-force conversion is the use of bolts or screws for joining parts together. In Fig. 7.30, a bolt and a nut are tightened to clamp two members together. After tightening, a force equilibrium is achieved. The force analysis shows that the tensile force on the bolt is the same as the clamping force on the two plates. This clamping force is a conversion from the torque applied for tightening.

7.2.5 Torque amplification

There are many situations when a very large amount of torque is needed. One can accomplish this by using extra extension of a wrench. One useful tool is called impact wrench, which can apply a huge amount of torque without a long leverage. The principle of an impact wrench is beyond the scope of Statics and is related to the conservation of momentum in Dynamics. There are other options to apply huge torques. We will discuss many methods to amplify a torque.

The simplest way of torque amplification is through a gear pair. As shown in Fig. 7.31, two gears with different radii are engaged to transmit power. If we draw a free-body diagram for each gear, we will see that at the engaging point, the tangential force, $\vec{F}$, and the normal force, $\vec{N}$, are the same for each gear. If the smaller gear in Fig. 7.31 is the driving gear and a torque, $\vec{\tau}_1$, is applied to it, the resulting tangential force can be determined by $\tau_1 = R_1 \times F$, assuming that the gear is running at a constant speed. For the driven gear, the resulting torque, $\vec{\tau}_2$, is induced by the same tangential force but with a larger radius, thus, $\tau_2 = R_2 \times F$. As a result, a torque amplification, $\frac{\tau_2}{\tau_1} = \frac{R_2}{R_1}$, is accomplished.

The cost of this torque amplification is that the rotation speed of the driven gear is reduced by the same factor. There are many different gear train arrangements for the purposes of torque amplification or speed

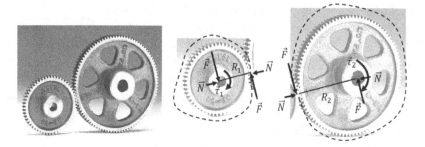

Figure 7.31: A gear pair as the example of torque multiplication.
Source: http://4.bp.blogspot.com/-Q4swrjvWSM0/UdOozSzou7I/AAAAAAAAJLU/w
Mp8cmciyCg/s476/SPUR+GEARS+001.jpg.

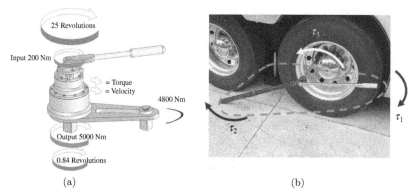

(a)   (b)

Figure 7.32: The work principle of a torque multiplier and its application to tighten (or loosen) a tire lug nut: (a) operation of a torque multiplier and (b) FBD for the torque multiplier.
Source: (a) http://www.powermaster.in/en/bolting-tools/images/dwg_1.jpg; (b) http://
image.trucktrend.com/f/48799656+re0+cr1+ar0/1304rv-03%2btorque-multipliers-field
-tested%2btorque-bar-and-torque-wrench.jpg.

reduction. A famous gear train is called planetary gear set. We will not discuss how a planetary gear set works in this book. A planetary gear set is like a mechanical calculator. It can combine two rotational inputs into one rotational output.

A special hand tool, called torque multiplier, utilizes such a planetary gear set for torque amplification. Such a device is shown in Fig. 7.32.

As shown in Fig. 7.32, if we apply $\tau_1 = 200$ Nm to the torque multiplier, the output torque will be amplified to $\tau_2 = 5000$ Nm, a factor of 25. However, it will take 25 revolutions of the input lever to achieve less than one revolution of the output shaft. Note that this large torque cannot be created out of nowhere. There is an extra lever on the torque multiplier to give the additional torque of 4800 Nm needed to achieve the 5000 Nm output torque. The use of such a torque multiplier for tightening the lug nut of a large truck tire is shown in Fig. 7.32(b). In this case, the additional torque is supplied by the contact force from the ground and $\tau_3 = \tau_1 + \tau_2$.

7.2.6 Torque wrench

In the previous section, we discussed the amplification of torque. However, one critical issue in assembling is to avoid over- or under-torquing. There are many designs of torque wrenches to allow a user to apply the correct torque. With the miniaturized electronic devices available today, an electronic wrench with sensors (i.e. strain gauges) can be made to precisely measure and display the torque during the application. An alarm can be raised when the desired torque value has been achieved. However, mechanical-type torque wrenches are still in wide use and provide a reliable and easy way to set correct torques, at a lower cost. One particular mechanical torque wrench, called click-type, is shown in Fig. 7.33(a).[3]

A torque setting is typically set through turning the handle in a click-type torque wrench. When the preset value of the torque has been reached, a click sound can be heard and a "skidding" feeling of the wrench can be felt so that the user could stop to avoid over-torquing.

The inner working mechanism of the torque wrench is shown in Fig. 7.33(b). In Fig. (7.33(c)), the key elements of the torque preset-ting mechanism are shown. The overall free-body diagram is shown in Fig. 7.33(a). A force $\vec{F}$ is applied by the user, which produces a torque $\tau = FL$. Inside the torque wrench, the wrench head is connected to a rod, G, pinned at point C to the torque wrench top tube D. If we draw a boundary to isolate element G and the wrench head, as shown in Fig. 7.33(c), there is one contact point (H) and two constraints at head center O and pin C. We only assign a normal force, $\vec{N}$, for the contact point (H) between element G and the ball (B) by ignoring the friction force. If the applied torque makes rod G rotate at an angle, α, then the contact point between rod G and ball B would shift to point H. At static equilibrium, the relationship between force $\vec{N}$ and torque, τ, can be established by constructing a moment equation

---

[3]US Patent US7451674B2. Adjustable Click Type Torque Wrench.

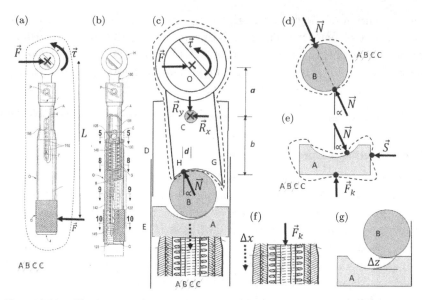

Figure 7.33: The inner working mechanism and force analysis of a click-type torque wrench: (a) FBD for a torque wrench; (b) internal design of the torque wrench; (c–g) FBDs for the internal elements.

Source: (a–b) see footnote 3

with respect to point C,

$$\sum (\vec{\tau})_C = 0: \quad -N\,d\cos(\alpha) - N\,b\sin(\alpha) - F\,a + \tau = 0 \qquad (7.30)$$

Introducing $\tau = FL$, we have

$$N = \frac{F\,(L-a)}{b\,\sin(\alpha) + d\cos(\alpha)} \qquad (7.31)$$

According to Fig. 7.31(d–f), force $\vec{N}$ also pushes down element A by compressing the spring below. Force $\vec{N}$, the spring force, $\vec{F}_k$, and the compression of the spring, Δx, are related as follows:

$$- N\cos(\alpha) + F_k = 0 \qquad (7.32)$$

$$F_k = k\,\Delta x \qquad (7.33)$$

where k is the spring constant.

We have

$$\Delta x = \frac{N\cos(\alpha)}{k} \qquad (7.34)$$

Note that Δx is the total spring compression. The torque wrench is designed so that when element A has a displacement of Δz, ball B will shift out of the concave surface of element A. Note that Δx is not equal to the actual displacement of element A because the spring already has a compression before the torque was applied. This pre-compression is achieved by rotating the handle element E to set the desired torque. Let us denote this pre-compression as Δw. When $\Delta x = \Delta w + \Delta z$, element A will move down by Δz, which makes ball B shift out of element A, as shown in Fig. 7.33(g). From there, ball B is free to move further to the right, and it would hit the wall of tube D, generating a clicking sound, and the skidding feeling. By setting the amount of Δw, through rotating tube E, the torque needed to cause the movement of Δz can be set, thus achieving the function of a torque wrench.

This mechanism of using a ball and two concave surfaces (elements G and A) for force setting can be found in many designs.

7.2.7 Torque application

Precise torque applications are critical in mechanical assembling practice. However, correct torque applications are often misunderstood or misused. In fact, the force mechanism and parameters involved in the torque-to-force conversion are different from case to case. There is an industry dedicated to torque application and calibration.

In automotive applications, the required torques are recommended in repair manuals. For example, a torque in the range of 15–21 Nm is often recommended for tightening parts of an aluminum engine. The torque for tire/wheel lug bolts is about 110–120 Nm. The tightening torque for the crankshaft pulley/harmonic balancer for a Mercedes M104 engine could be 200 Nm, plus an additional 90-degree turn. As a result, the actual torque is much higher but not explicitly specified.

Another highly critical example of torque applications is the tightening of head bolts for clamping the cylinder head to the engine block of an internal combustion engine. For Mercedes M104 engines, the following tightening sequences are specified in the repair manual:

"*Threads of cylinder head bolts and contact surfaces of washers*
. . . oil with engine oil."

"*Cylinder head bolts . . . Tighten in stages in the order of the tightening diagram, starting with 1.*"

"1ˢᵗ stage 55 Nm"

"2ⁿᵈ stage 90 degree tightening angle"

"3ʳᵈ stage 90 degree tightening angle"

The above description illustrates a torque practice, called wet torque-ing, when lubricants are used on the bolt's threads. Again, the final torque is not specified explicitly. The head bolt tightening is highly critical because poor application could lead to a blown head gasket.

The term, "wet torqueing", is converse to the typical "dry torqueing", for which no lubricant is used between the bolt and the bearing surface. This observation points to the important role of friction in the torque application.

The following relationship formula between the clamping force and the torque is suggested in *Tohnichi Torque Handbook, Volume 7*,[4]

$$\tau = R \left\{ \frac{d_2}{2} \left(\frac{\mu}{\cos \alpha} + \tan \beta \right) + \mu_n \frac{d_n}{2} \right/ 1000 \tag{7.35}$$

where τ is the tightening torque in Nm; R is the tensile force on the bolt or the clamping force in N, d_2 is the pitch diameter in mm of the bolt, μ is the friction coefficient of the bolt threads, α is the half angle of the screw thread, β is the lead angle of the thread, μ_n is the friction coefficient between the bolt head and the bearing surface, and d_n is the pitch diameter of the bearing surface in mm.

For an M8 hex bolt, the following parameter values can be defined: $d_2 = 7.188 \, \text{mm}$, $\mu = 0.15$, $\alpha = 30°$, $\mu_n = 0.15$, $d_n = 11.27 \, \text{mm}$, and $tan\beta = 0.0554$. If we want to have a clamping force $R = 8000 \, \text{N}$, the tightening torque can then be calculated as $\tau = 13.4 \, \text{Nm}$. Typically, such a tightening torque will be rounded to $15 \, \text{Nm}$. Although the formula of Equation (7.35) is useful and generally reliable, there is a serious drawback related to the assumption of the friction coefficients, μ and μ_n. The surface conditions of the bolt and the nut, such as finish, rustiness, greasiness, or cleanliness, can change the values of the friction coefficients significantly. If the practice is "wet torqueing", the friction coefficient will be totally different. In addition, as the clamping force increases, the friction can change, too. Therefore, Equation (7.35) is often used as a reference, and more studies or measurements will be needed to verify if the actual clamping force is achieved when the application is highly critical.

Another question is how the clamping force is determined. One simple practice is to set the clamping force as high as possible without causing

---

[4] *A Guide to Torque Management and Tohnichi Products*, ASIN: B01N4CLNYD. Available from http://tohnichi.jp.

plastic deformation or yielding of the bolt, combined with a safety factor. For example, common practice is to set the clamping force to be 70% of the yield strength multiplying the load bearing area of the bolt. If this force is not sufficient, then multiple bolts are used.

To tighten a bolt at a specified torque, a torque wrench is used, as discussed in Section 7.2.6. Today, one can buy an electronic torque gauge and convert any wrench into an electronic torque wrench. Using a torque wrench helps to prevent over-tightening. The thread could be stripped, in particular when tightening a bolt to the tapped hole of an aluminum body.

In most tire shops, torque wrenches are used to avoid over-tightening, which could lead to damage to the brake rotor and the wheels. Furthermore, when lug bolts are over-tightened, the driver might not be able to release the bolts for changing the tires. In some tire shops, the practice of using the torque wrench could be wrong. Often, a technician would use an impact wrench to tighten the lug bolts to save time, followed by using the torque wrench to verify the torque. However, if the technician is too aggressive in using the impact wrench, the bolt will be over-tightened already. The subsequent use of the torque wrench is then only to confirm that the actual tightening torque is no less than the specified torque. Correct practice is to use the impact wrench very conservatively and only tighten the bolts to less than half the specified torque. A torque wrench is then used to tighten the bolt in a few steps until the specified torque is reached.

In production lines, torque wrenches with fixed torque settings are often used to ensure consistency in the assembly. Torque setting, calibration, and application are critical practices in factories.

7.3 Fixtures

A fixture is the arrangement of constraints and contacts to hold an object at equilibrium against external forces and moments. A fixture can be well-constrained or over-constrained, but must not be under-constrained. If we calculate the degrees of freedom of an object based on Gruebler's equation, a well-constrained system has zero degrees of freedom, an over-constrained system has a negative degree of freedom, while an under-constrained system has a positive degree of freedom.

7.3.1 Fixtures with pined–pined rod constraints

If we only use pined–pined rods as constraints for fixtures, we can identify the constraint conditions of a 2D body clearly as shown in Fig. 7.34.

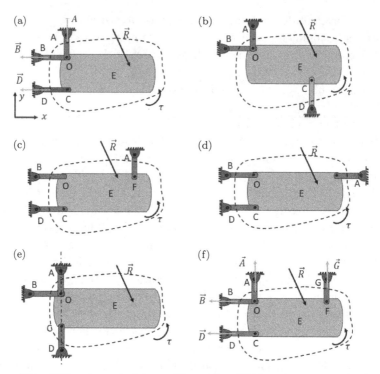

Figure 7.34: Different constraint conditions of a 2D fixturing system: (a–c) well-constrained systems; (d–e) under-constrained systems; (f) over-constrained system.

Figures 7.34(a)–(c) are well-constrained systems, Figs. 7.34(d) and 7.34(e) are under-constrained systems, while Fig. 7.34(f) is an over-constrained system. For Fig. 7.34(a), we can write down three equilibrium equations based on the free-body analysis,

$$\sum F_x = 0: -B - D + B_x = 0 \tag{7.36}$$

$$\sum F_y = 0: A - B_y = 0 \tag{7.37}$$

$$\sum (\vec{\tau})_O = 0: -\tau_D - \tau_R + \tau = 0 \tag{7.38}$$

where τ_D and τ_R are moments by forces $\vec{D}$ and $\vec{R}$ with respect to point A. Based on these three equilibrium equations and given any external loadings of $\vec{R}$ and $\vec{\tau}$, we can determine the corresponding values of the three constraint forces, $\vec{A}$, $\vec{B}$, and $\vec{D}$, uniquely. As a result, the rigid body E can maintain its position under loadings. Because we can determine uniquely the constraint forces, we can determine if these rods are strong enough for the task. A similar analysis can be applied to Figs. 7.34(b) and 7.34(c).

As discussed before, Gruebler's equation is not reliable in determining the degrees of freedom for Figs. 7.34(d) and 7.34(e). For Fig. 7.34(d), there are no constraint forces to resist any external force in the y direction because all three constraint forces are in the x direction. For Fig. 7.34(e), the three constraint forces cannot resist a moment with respect to point O.

For Fig. 7.34(f), the three equilibrium equations are

$$\sum F_x = 0: \ -B - D + B_x = 0 \tag{7.39}$$

$$\sum F_y = 0: \ A + G - B_y = 0 \tag{7.40}$$

$$\sum (\vec{\tau})_O = 0: \ -\tau_D - \tau_R + \tau_G + \tau = 0 \tag{7.41}$$

We have four unknowns, A, B, D, and G, but only three equations. As a result, the values of these constraint forces cannot be uniquely determined. The system is then over-constrained. Note that in reality, the values of these forces are still unique. It is only that we do not have enough equations to determine them. If one installs a sensor to measure one of the forces, then we can determine the remaining three forces uniquely. It is also possible to establish a fourth equation based on the geometry and elastic deformations, which will be addressed in Chapter 8.

Similarly, to fix a 3D body, we will need to use six pined–pined rod constraints. Figure 7.35 illustrates these conditions. Figures 7.35(a)–7.35(c) are well-constrained, Figures 7.35(d) and 7.35(e) are under-constrained, while Fig. 7.35(f) is over-constrained. Figure 7.35(d) is under-constrained because all constraint forces pass through the dashed line shown in the figure; thus, any moment around this line cannot be countered. The 3D fixturing is harder to visualize. For the well-constrained cases, it follows a 3-2-1 scheme, for which there are three constraints in the z direction over three points, two constraints in the x (or y) direction, and one constraint in the y (or x) direction. Another 3-2-1 scheme has three constraint forces at one point in three directions, two constraint forces in two directions at a different point, and one constraint force in one direction at a third point as in Fig. 7.35(a). It can also be a 2-2-2 scheme as in Fig. 7.35(b), and a 3-1-1-1 scheme as in Fig. 7.35(c).

A practical example of using the 2-2-2 scheme for fixturing is denoted as a hexapod, as shown in Fig. 7.36(a). If these constraints are replaced

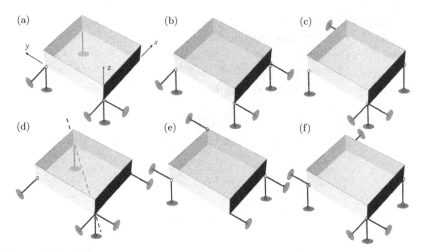

Figure 7.35: Different constraint conditions of a 3D fixturing system: (a–c) well-constrained system; (d–e) under-constrained systems; (f) over-constrained system.

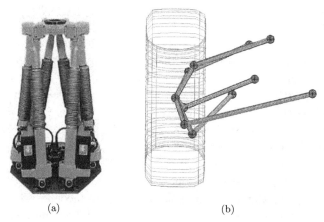

Figure 7.36: (a) A hexapod structure and (b) a five-link rear suspension system which only allows the up and down motion of the tire.

Source: (a) http://fennetic.net/gingery_machines/images/f200ia.jpg; (b) https://upload. wikimedia.org/wikipedia/commons/9/92/5link3Drear1.gif.

by hydraulic cylinders, then the position of the object can be controlled to different positions within the range.

One can also design a fixturing system with the pined–pined rod constraints so that only one or two specific degrees of freedom can be allowed. Figure 7.36(b) illustrates such a system, which is used in the rear suspension of a car to allow for only the up and down motion.

7.3.2 Fixturing through point contacts — kinematic coupling design

The fixturing system discussed in Section 7.3.1 based on pined–pined rod constraints is deterministic but is also permanent. In manufacturing, parts need to be held in place precisely under loading. The setup and removal of the part should also be quick and easy, for which the pined–pined rod constraint is not suitable.

Recalling the point contact discussed in Section 6.1.1, the contact force at the contact point is directional along the normal direction, as shown in Fig. 6.3 in Chapter 6. If we do not allow the object to have a tendency of movement in the tangential direction, the frictional force can be ignored. As a result, a point contact can be an effective constraint for fixturing because it offers a precisely defined force. Utilizing point contacts for fixturing to achieve a deterministic constraint system is called kinematic coupling design, as discussed in Slocum (1992).[5]

With the point contact for kinematic coupling design, each contact point provides a constraint force in the normal direction. Note that this force can only be compressive. There will be additional clamping forces to keep the object from losing contact with the contact points. These clamping forces will be addressed later. Let us focus on how to arrange the contact points. The point contact will be achieved by pairing a ball with a flat surface.

Figure 7.37 illustrates the 3-2-1 scheme. A cone cavity and a ball achieve three contact points, a v-groove and a ball achieve two contact points, and a flat surface and a ball, one contact point. We can also use the 2-2-2 scheme, for which three pairs of balls and v-grooves are used. Figure 7.38 illustrates how two rigid disks can be positioned with respect to each other quickly, repeatedly, and precisely using the 3-2-1 scheme or the 2-2-2 scheme.

In general, the 2-2-2 scheme is preferred because it is harder to create the cone cavity than the v-groove. As shown in Fig. 7.39, once the plates are clamped by a clamping force, $\vec{R}$, these two disks will be fixed securely as one.

---

[5]Slocum, A.H. (1992). *Precision Machine Design.* Prentice Hall Inc. ISBN 0-13-690918-3.

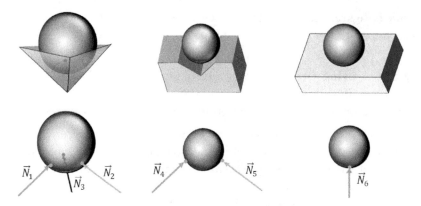

Figure 7.37:   The 3-2-1 kinematic coupling design with point contacts.

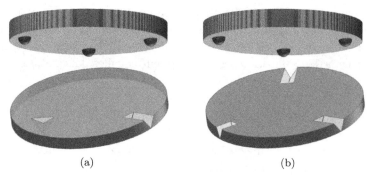

<table>
<tr><td>(a)</td><td>(b)</td></tr>
</table>

Figure 7.38:   Kinematic coupling designs for positioning two disks precisely with respect to each other: (a) the 3-2-1 system and (b) the 2-2-2 system.

The advantage of this kinematic coupling design for fixturing is that the repeatability is very high. As reported in Slocum (1991), sub-micron positioning accuracy can be maintained after over 500 cycles of mounting and dismounting if wear-resistant materials are chosen for the ball and the v-groove.

7.3.3 Fixturing through surface and point contacts

Although kinematic coupling design is elegant and effective, it is not suitable for general fixturing applications when parts are not equipped with the balls and v-grooves. The 3-2-1 scheme can still be used in general

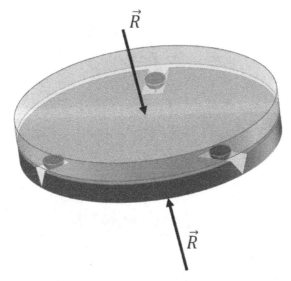

Figure 7.39: Two plates are clamped together precisely via the use of the kinematic coupling design.

fixturing practice. The three contact points establish a plane, the two contact points establish a line, and the last contact point acts as stop. Figure (7.40) illustrates how this general 3-2-1 scheme works.

To place a part correctly onto the fixture, one should first align the bottom surface of the part with the flat surface of the fixture (surface A), which theoretically is established by three points. However, we really do not know which three points are used exactly. Assuming that the bottom surface of the part is sufficiently flat, then the part can only move over the fixture surface, restricting to only three degrees of freedom. Then, the part should be moved, while maintaining the contact with surface A, until one side of the part is in contact with the posts B and C. The two contact points on the posts establish a line. The part should then move along this line (maintaining contacts with posts B and C), while still in contact with surface A, until it is stopped by post D. After that, a clamping force can be applied to hold the part securely in place. The fixturing setup of Fig. 7.41 is widely used and specified in ASME Y14.5.1M: Geometric Dimensioning and Tolerancing (GD&T).

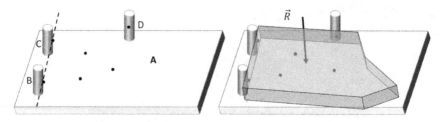

Figure 7.40:   General fixturing setup using the 3-2-1 scheme.

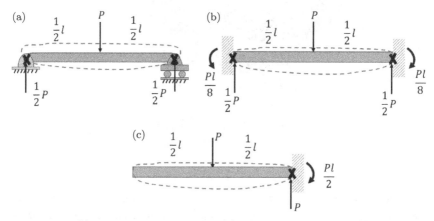

Figure 7.41:   (a) A simply supported beam; (b) a doubly clamped beam; (c) a cantilever beam.

7.4 Metal and Wood Structures

The structures of a machine need to be held without any degrees of freedom; however, there is no need to stay with a well-constrained system. Typically, the structure is an over-constrained system for additional structural rigidity. The truss system we discussed in Section 7.1.3 is an example of machine structures, but it is rarely used in machines. We will discuss several rigid structures in the following sections. In the truss system, pins are used to join elements together. In machine structures, total constraints are typically used. Figure 7.41 illustrates the main difference between a clamped support and a pin support.

For the simply supported beam with pin constraints as in Fig. 7.41(a), the pin supports only provide vertical constraint forces without moments. On the other hand, for a doubly clamped beam (Fig. 7.41(b)), there are constraint moments and forces at the clamped supports. The free-body

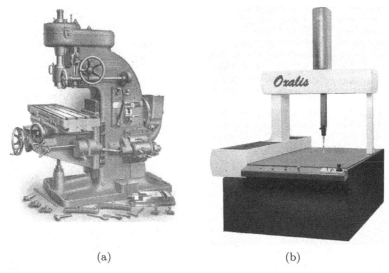

(a)   (b)

Figure 7.42: Two different machine structures: (a) an open configuration and (b) a closed configuration.

Source: (a) https://upload.wikimedia.org/wikipedia/commons/a/a2/Practical_Treatise _on_Milling_and_Milling_Machines_p016.png; (b) https://upload.wikimedia.org/wikiped ia/commons/b/bf/Macchina_di_misura_3D.jpg.

diagram analysis we have discussed so far in this book does not provide the solution to determine the size of the moment shown in Fig. 7.41, which would require an additional equation related to the boundary condition of the clamped support, to be discussed in Chapter 8. The moment here is called the bending moment, which directly contributes to the normal stress of the beam, to cause failure. The maximum bending moment also resides at the point the beam is clamped; so do the failures. If the beam is only clamped at one side, i.e. a cantilever beam, as shown in Fig. 7.41(c), the resulting moment at the clamped support is now four times larger. As a result, the cross-section of a cantilever beam must be large enough to prevent failures.

7.4.1 Rigid machine structure

For machines, the structure is designed based on many different factors. Figure 7.42 illustrates two different configuration concepts: (a) open configuration and (b) closed configuration.

An open configuration, which is similar to a cantilever beam, provides easier access, while a closed structure, which is similar to a doubly clamped beam, provides better structural rigidity. However, as discussed in the previous section, an open configuration could be subjected to higher bending moments, prone to deformations and failures. As a result, the machine structure in Fig. 7.42(a) is substantially thicker at the shoulder. However, it should be noted that the machine shown in Fig. 7.42(a) is for machining with a heavy loading condition, while the machine in Fig. 7.42(b) is for metrology, with light loading condition but high geometric stability and rigidity. The statement above is only for qualitative purposes.

7.4.2 Woodworking structure

As discussed in Section 7.4.1, a cantilever structure is subjected to substantial bending moment. Some structure designs, however, must have the cantilever design. One example is table legs. As shown in Fig. 7.43(a), a table leg, A, must withstand bending moments in two directions, $\vec{M}_x$ and $\vec{M}_y$, based on the free-body diagram which isolates the leg. Usually, two side pieces, B and C, bonded to the leg by two screws each, are used to resist these two moments. Each pair of the screws creates a force couple to resist the moment, i.e. force couple, $\vec{F}_x$ for $\vec{M}_y$ and $\vec{F}_y$ for $\vec{M}_x$. The side pieces should have proper width so that the distance of the force couple is sufficiently large to create effective moments.

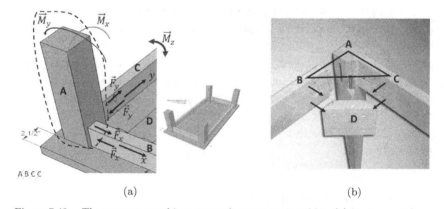

(a) (b)

Figure 7.43: The support to achieve strong leg support in tables: (a) leg support for a table with a fixed top and (b) leg support for a table with a removable top.

Source: (a) https://i.stack.imgur.com/RhnwX.jpg; (b) https://i.stack.imgur.com/r4L3K.jpg.

This fixturing design of the table has a weakness in how the side piece is held. For example, piece C is susceptible to moment $\vec{M}_z$, as shown in Fig. 7.43(a). To resist $\vec{M}_z$, a table top is needed. The side piece needs to be bonded to the table top to secure the side pieces. This design, however, is not suitable for a table with a removable top, such as a glass top.

An alternative design is shown in Fig. 7.43(b), where a third piece (D) is introduced so that the leg (A), the side pieces (B and C), and the additional piece (D) form a strong corner, just like a basic triangular truss system.

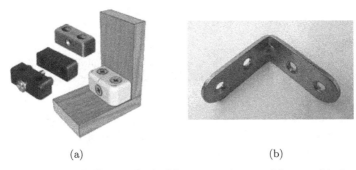

(a)   (b)

Figure 7.44:   Off-the-shelf pieces for building strong corners: (a) corner blocks and (b) an angle bracket.

Source: (a) http://wiki.dtonline.org/images/thumb/f/f0/ModestyBlocks.png/200px-Mo destyBlocks.png; (b) https://upload.wikimedia.org/wikipedia/commons/thumb/0/0a/ Bronze_angle_bracket_-_4_x_4_cm_-_A.jpg/1200px-Bronze_angle_bracket_-_4_x_4_cm_-_A. jpg.

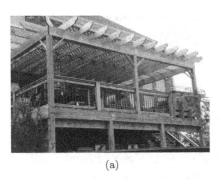
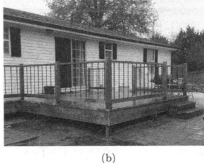

(a)   (b)

Figure 7.45:   (a) Top deck post with strong corners and (b) deck posts without strong corners.

Source: (a) https://www.polkadotchair.com/wp-content/uploads/2010/07/deckafter1. jpg; (b) https://i.stack.imgur.com/8M7FS.jpg.

(a)   (b)

Figure 7.46:   (a) Bookshelf without strong corners and (b) bookshelf with strong corners with back board.
Source: (a) http://3.design-milk.com/images/2013/04/tensile-plywood-bookshelf-des ign-soil-600x397.jpg; (b) https://1.bp.blogspot.com/_tfGC7tOlrdk/TTQCBjZ08Cl/AA AAAAAALEc/-zxo4F68GaQ/s1600/TrueModern-low-bookcase-design-public.png.

From Fig. 7.43(b), we realize that an important practice in furniture making is related to building strong corners. There are commercial angle pieces for building strong corners, as shown in Fig. 7.44. In deck building, large posts are established and then railings are built, as shown in Fig. 7.45. Additional pieces can be seen in the deck shown in Fig. 7.45(a) for stronger corners. However, in some cases, such as the deck in Fig. 7.45(b), the posts are mounted only with nails at the bottom. As the wood ages, it might crack and the posts will no longer be held securely. It could be dangerous when a person leans on the railing with such under-supported posts, especially for a high deck. The entire railing could fall over, causing injuries. Angle pieces such as those shown in Fig. 7.44 are useful to secure these deck posts.

Strong corners can also be formed by using a back plate, similar to the table top, as shown in Fig. 7.46(b). On the other hand, the shelf shown in Fig. 7.46(a) is weak when it is pushed from the side.

7.5 Concluding Remarks

In this chapter, we apply the free-body diagram analysis for many practical applications. We first discuss how to use constraints to eliminate the movement of an object, which leads to the discussion of truss systems and their extensions. The force analysis related to hand tools is then presented, with focus on how the tools are used to achieve force/torque conversions and force/torque amplifications. The principle of kinematic coupling for fixturing is then illustrated, followed by the discussion of rigid machine structures and woodworking structures. All the above analyses are conducted with the assumption of rigid bodies. In the next chapter, we will present force analysis for non-rigid bodies and for situations when a system is over-constrained.

Chapter 8

What if the Free-Body Diagram Is Not Enough?

In Chapter 7, we demonstrated how to use the free-body diagram to conduct correct force analysis for many practical applications. It appears that we were able to address all those problems satisfactorily using the free-body diagram. However, the free-body diagram has its limitations. First, there is the assumption of rigid bodies. Second, sometimes, we must construct many free-body diagrams so that we have enough equations to solve for the unknowns. Third, in some cases, we simply cannot solve for all the unknown forces/moments no matter how many free-body diagrams we have constructed. This situation happens in over-constrained systems. Fourth, a system might be so complicated that the construction of free-body diagrams becomes very difficult and not really useful. Fifth, a free-body diagram always indicates the effect of external forces on the free-body. Within the free-body, there are still forces, denoted as internal forces, which cannot be represented by the free-body diagram. The consideration of the internal forces could be important in many situations. In other words, as powerful as it is, the free-body diagram has its limits. As the title of this chapter suggests: What if the free-body diagram is not enough?

We will discuss the limitations of the free-body diagram in this chapter and introduce some new and useful techniques.

8.1 What Do We Do if the Bodies are not Rigid?

A rigid body, as suggested by its name, is a body that does not deform when it is subjected to forces and moments. Of course, as we discussed in Chapter 2, the entire free-body can be considered as rigid if an equilibrium has been achieved without additional deformations. However, from the initial state to the equilibrium state, the free-body may undergo deformations if it is not rigid. How should this deformation be considered and how could it be useful in solving problems? This is the main focus of this section.

8.1.1 Types of deformation

There are two types of deformations, denoted as the elastic deformation and the plastic deformation. In an elastic deformation, a body can return to its original shape once the forces/moments are removed from the body. On the other hand, if the body cannot return to its original shape, it indicates a permanent deformation has occurred. This deformation is then called plastic deformation.

For most of the systems we designed, we do not want plastic deformations during operation. For example, a tennis racket, upon hitting a ball, can deform but will return to its original shape after hitting. When a tennis player loses his/her temper and slams the racket to the ground, the racket would break or distort, losing its original shape, no longer usable. On the other hand, in many engineering applications, we actually want plastic deformation. For example, the beautiful curvatures of automobile body panels are fabricated from a flat sheet metal through plastic deformation.

In this section, we will limit ourselves to the cases of elastic deformation. The topic of plastic deformation is beyond the scope of this book.

8.1.2 Elastic deformation

The most obvious example of the elastic deformation is related to linear springs as in Fig. 8.1(a). As shown in Fig. 8.1(b), when a linear spring is attached to an object (A) and a force $\vec{P}$ is applied to the object, the object will move to the right, stretching the spring. As a result, the spring applies a force, $\vec{F}_k$, to resist the movement. In general, the forces $\vec{P}$ and $\vec{F}_k$ are different unless object A is at equilibrium. Based on Hooke's law for linear springs, we can establish a relationship between the spring force and its deformation,

$$F_k = k \left(l - l_o \right) = k \, \Delta l \tag{8.1}$$

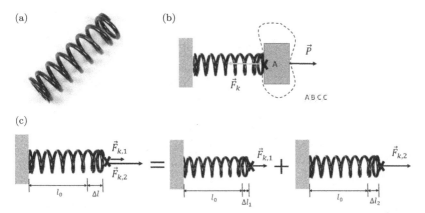

Figure 8.1:   (a) A linear spring; (b) a linear spring attached to an object under loading; (c) the principle of superposition applied to the linear spring.

Source: (a) https://i.stack.imgur.com/Z1MQb.jpg.

where k is the spring constant with a unit of $\frac{N}{m}$ or lb_f/ft, l_o is the original length of the spring before any force is applied, l is the new length after the force is applied, and Δl is the deformation of the spring. Equation (8.1) is the linear relationship between the force and the deformation of a linear spring. A positive deformation is when the spring is stretched, while a negative deformation is that when the spring is being compressed. The relationship is linear because if two forces are applied to the spring, then the combined deformation is equal to the deformation of each force combined. Therefore,

$$F_k = F_{k,1} + F_{k,2} = k\ \Delta l = k\ \Delta l_1 + k\ \Delta l_2 \tag{8.2}$$

Based on the principle of superposition, when the spring is subjected to multiple forces, we can apply each force individually and then add up each corresponding deformation to find the total deformation, as shown in Fig. 8.1(c). The principle of superposition is a very powerful tool to reduce a complicated problem to simpler cases as long as the linear relationship between the force and the deformation is maintained.

Similar to linear springs, we also have torsional springs, for which a moment is applied to cause a torsional deformation, as shown in Fig. 8.2. A similar linear relationship exists between the moment and the torsional deformation, and the principle of superposition still applies. Torsional springs are widely used in mechanical watches.

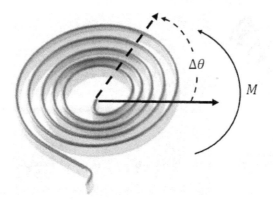

Figure 8.2: The deformation of a torsional spring.
Source: http://i.stack.imgur.com/7FTEK.jpg.

Figure 8.3: An elastic mechanical element subjected to an axial force is a linear spring with a stiff spring constant.

8.1.3 Mechanical elements as linear springs

The actual mechanical elements used for structures are typically strong enough to be considered as rigid bodies. However, these elements still demonstrate elastic deformations even though they could be small. For example, a steel bar subjected to an axial force, as shown in Fig. 8.3, will have a very small deformation, ΔL, if the force is not too high to cause plastic deformation. The relationship between the deformation and the force can be defined as, as long as $\Delta L \ll L_0$,

$$\Delta L = F_k L_0 / (EA) \tag{8.3}$$

where E is Young's modulus of the material and A is the cross-section area of the bar.

Rewriting Equation (8.3), we have

$$F_k = \frac{EA}{L_0} \Delta L = k \Delta L \tag{8.4}$$

The spring constant of this mechanical bar is then defined as $k = \frac{EA}{L_0}$. We can see that a thinner bar will have a lower spring constant, so is a longer bar. A steel bar will have a higher Young's modulus than a wood bar to achieve a higher spring constant.

8.1.4 First example when FBD is not enough

Now, consider a bar supported by two rigid ends, as shown in Fig. 8.4, and subjected to an axial force.

We can construct an FBD as shown in Fig. 8.4(a) by considering the bar at equilibrium as a rigid body; then we have

$$F - R_A - R_B = 0 \tag{8.5}$$

We have two unknowns but only one equation; therefore, we are unable to determine the values of R_A and R_B. By considering the bar as an elastic body, or a 1D spring, we will be able to find a second equation needed to solve the problem.

As shown in Fig. 8.4(b), we first replace the fixed end at point B by its equivalent constraint force, R_B. Then, we can invoke the principle of superposition by considering the effect of force F and R_B, separately, as shown in Fig. 8.4(c).

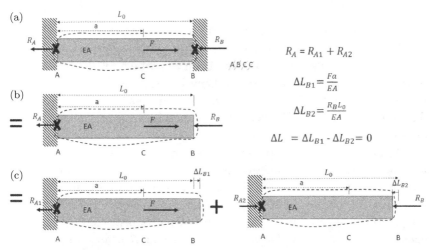

$$R_A = R_{A1} + R_{A2}$$

$$\Delta L_{B1} = \frac{Fa}{EA}$$

$$\Delta L_{B2} = \frac{R_B L_0}{EA}$$

$$\Delta L = \Delta L_{B1} - \Delta L_{B2} = 0$$

Figure 8.4: (a) An elastic mechanical element fixed at both end, subjected to an axial force; (b) FBD with one clamp removed; (c) FBDs using superposition.

Based on Equation (8.4), we have

$$\Delta L_{B1} = \frac{Fa}{EA} \text{ and } \Delta L_{B2} = \frac{R_B L_0}{EA} \tag{8.6}$$

where ΔL_{B1} is the deformation due to force F, while ΔL_{B2} is due to R_B. Because in the original condition, there is no deformation at point B, due to the fixed end, the total bar deformation,

$$\Delta L = \Delta L_{B1} - \Delta L_{B2} = 0 \tag{8.7}$$

Equation (8.7) is the second equation we need to solve for R_A and R_B, combined with Equation (8.5). We found

$$R_B = \frac{Fa}{L_0} \text{ and } R_A = \frac{F(L_0 - a)}{L_0} \tag{8.8}$$

Note that $R_A = R_{A1} - R_{A2}$ in Fig. 8.4(c).

8.1.5 Second example when FBD is not enough

The deformation of a bar could be due to temperature changes, denoted as thermal expansions. For a straight bar, the axial thermal expansion can be expressed as

$$\Delta L = \alpha L_0 \Delta T \tag{8.9}$$

where α is the thermal expansion coefficient. It is clear that a longer bar will have a larger thermal expansion for the same temperature increase. Accommodating thermal expansions is an important task in mechanical design. We will not get into the mechanical design issues here, but as we discussed in Chapter 7 (Fig. 7.8), a truss system should be supported by a fixed pin and a pin on wheels at the foundation to accommodate thermal expansions.

If a bar is supported by two fixed ends, not allowing thermal expansions, as shown in Fig. 8.4, thermal expansion can cause thermally induced loads on the bar. As shown in Fig. 8.5(a), thermally induced loads, R_A and R_B, could be induced due to the temperature increase, ΔT. When the temperature rises, the bar needs to expand but is restricted by the fixed ends; as a result, the constraint forces, R_A and R_B, are induced. This is similar to a person wearing a tight pair of jeans and gaining weight. The jeans become tighter. However, if we do an FBD, as in Fig. 8.5(a), we

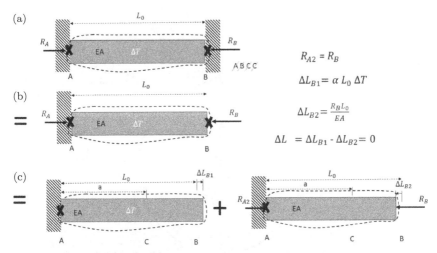

Figure 8.5: (a) An elastic mechanical element fixed at both ends and subjected to a thermal expansion; (b) FBD with one clamp removed; (c) FBDs using superposition.

have

$$R_A - R_B = 0 \qquad (8.10)$$

Equation (8.10) contains two unknowns and we have $R_A = R_B$, but we do not know the value. Now, using the same approach as described in Fig. 8.4, we have, as shown in Figs. 8.5(b–c),

$$\Delta L_{B1} = \alpha L_0 \Delta T \text{ and } \Delta L_{B2} = \frac{R_B L_0}{EA} \qquad (8.11)$$

With the same argument as with Equation (8.7), we have

$$\Delta L = \Delta L_{B1} - \Delta L_{B2} = 0 \qquad (8.12)$$

Solving Equations (8.10) and (8.12), we have

$$R_A = EA\,\alpha\,\Delta T = R_{A2} = R_B \qquad (8.13)$$

8.1.6 Practical thermal expansion problems in bearings

Problems related to thermal expansions are critical to many industrial applications. One of them is the thermally induced preload in spindle bearings. Bearing preload is the internal compressive force between the bearing races and the rolling elements. Preload is set during installation; thus, the name PRE-load. This force removes the internal play of the

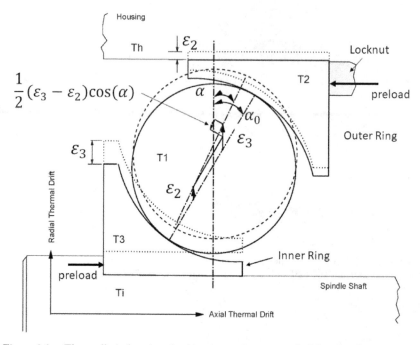

Figure 8.6: Thermally induced preload in an angular contact ball bearing due to uneven thermal expansions of the shaft, inner ring, ball, outer ring, and the housing. This figure is an improved image based on Stein and Tu (1994).

bearing and ensures the condition of rolling without slipping of the rolling element during operation. However, as the bearing heats up, the preload will change due to uneven thermal expansions of different bearing elements. This thermally induced bearing preload problem has been explored in several journal papers.[1] Similar to Equation (8.13), the thermally induced preload is related to temperature, but not just one temperature. Thermal expansions of different bearing elements in an angular contact ball bearing are shown in Fig. 8.6. The contact angle of the bearing is α_0 and the new angle is α after thermal expansions, while ε_1 is the thermal expansion of the ball after a temperature rise T_1, ε_2 is the thermal expansion of the outer

[1]Stein, J.L. and Tu, J.F. (1994). "A state-space model for monitoring thermally induced preload in anti-friction spindle bearings of high-speed machine tools". *Transactions American Society of Mechanical Engineers Journal of Dynamic Systems, Measurements, and Control.* **116**(3): 372–386.

Tu, J.F. (1995). "Thermoelastic instability monitoring for preventing spindle bearing seizure". *Tribology Transactions.* **38**(1): 11–18.

Tu, J.F. and Katter, J.G. (1996). "Bearing force monitoring in a three-shift production environment". *Tribology Transactions.* **39**(1): 201–207.

ring and housing after temperature rises T_2 and T_h, respectively, and ε_3 is the thermal expansion of the inner ring and shaft after temperature rises T_3 and T_i, respectively. The resultant thermal expansion that induces the thermal preload becomes

$$\varepsilon_t = \varepsilon_1 + \varepsilon_{10} + \frac{1}{2}(\varepsilon_3 - \varepsilon_2)\cos(\alpha) = \varepsilon_t(T_1, T_2, T_3, T_h, T_i) \qquad (8.14)$$

where ε_{10} is the compression due to the initial preload. The initial preload is achieved by tightening the locknut to force the outer ring against the ball, and subsequently against the inner ring, along the contact angle as shown in Fig. 8.6.

The term ε_t is similar to ΔL_{B1} in Equation (8.11) and is a function of the temperatures of different bearing elements. Because the bearing is not a straight bar as in the case of Fig. 8.5, the induced preload is not a linear function of ε_t, as in the case of Equation (8.1). The bearing is a nonlinear spring, and the thermally induced preload becomes

$$F_t = k_t \, \varepsilon_t^{1.5} \qquad (8.15)$$

where k_t is the stiffness constant of the bearing.

Equations (8.14) and (8.15) indicate that if the spindle housing is kept at a constant temperature without expansions, similar to the fixed end supports of Fig. 8.5, the thermally induced preload becomes higher. In other words, the thermal expansions of the housing and the outer ring are necessary to accommodate the thermal expansions of the ball and the inner ring to keep the thermally induced preload in check. The increased thermal preload will increase the friction loss of the bearing to raise the temperature further, which is a positive feedback mechanism, possibly leading to instability. Most of the time, the bearing can reach a thermal equilibrium and the induced thermal preload can be stabilized to a higher value without further increases. However, a thermal instability can occur if the thermal expansions of the inner ring and the ball outpace the thermal expansions of the outer ring and housing substantially. This is particularly true if the initial preload, which is related to ε_{10}, is excessive or the spindle speed is too high. In such cases, the thermally induced preload will continue to rise, causing the bearing to seize or fail catastrophically. Figure 8.7 is the measured thermally induced preload of a roller bearing spindle (see footnote 1) and thermal instability occurs when the spindle is running at 1400 rpm at time 1,000 second when the thermal preload kept increasing reaching over 7,000 N.

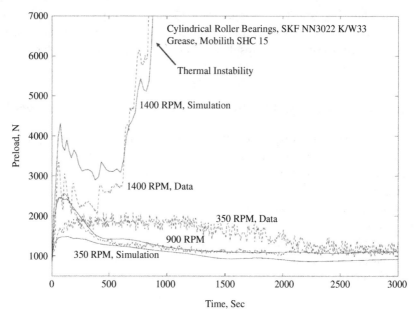

Figure 8.7: Measured thermally induced preload in a roller bearing. At 1400 RPM, thermal instability occurred at time 1,000 when the thermal preload kept increasing, reaching over 7,000 N. This figure is an improved image based on Stein and Tu (1994).

A similar instability problem can occur to the wheel bearings of cars. To avoid excessive preload, the locknut of the wheel bearings are usually only tightened by fingers (finger tight); otherwise, the wheel could seize during driving.

8.1.7 2D springs

In fact, every mechanical element can be treated as a spring. Some elements might have much higher spring constants (steel bars), and some lower (rubber bands). From experiments, we can easily establish the spring constant of an element. Sometimes, if the element has a simple geometry, we can derive the spring constant analytically. A flexible cantilever beam is a useful spring element for many applications, for example, as the leaf spring suspension used in vehicles, as shown in Fig. 8.8.

In fact, a cantilever beam is more than a linear spring. As the beam deforms by a linear force, the beam also demonstrates torsional deformation similar to a torsional spring, as shown in Fig. 8.9.

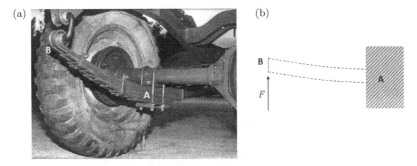

Figure 8.8: (a) A leaf spring suspension system and (b) an equivalent cantilever beam. *Source*: (a) https://upload.wikimedia.org/wikipedia/commons/6/63/Leafs1.jpg.

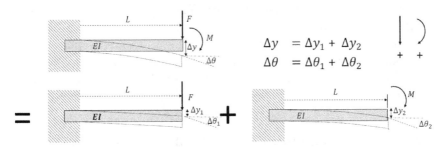

Figure 8.9: A cantilever beam is a 2D spring, and the principle of superposition applies.

We call an element similar to a cantilever beam as a 2D spring, and the governing equation (Hooke's law) is in a matrix form. In Fig. 8.9, we defined downward direction and clockwise direction as positive. The 2D spring equation becomes

$$\begin{bmatrix} F \\ M \end{bmatrix} = \begin{bmatrix} k_{11} & k_{12} \\ k_{21} & k_{22} \end{bmatrix} \begin{bmatrix} \Delta y \\ \Delta \theta \end{bmatrix} = K \begin{bmatrix} \Delta y \\ \Delta \theta \end{bmatrix} \qquad (8.16)$$

where K is the spring matrix or 2D spring constants.

In fact, we know $k_{12} = k_{21}$ due to the principle of reciprocity. We will not discuss this theorem here. The elements of k_{ij} are called cross-coupling spring constants. If we know the beam properties, with E as the elastic modulus and I as the second moment of area of the beam's cross-section, all the elements in the stiffness matrix can be defined as

$$k_{11} = \frac{12\,EI}{L^3}; \quad k_{12} = k_{21} = \frac{-6\,EI}{L^2}; \quad k_{22} = \frac{4\,EI}{L} \qquad (8.17)$$

Equation (8.16) can also be expressed in its compliance form,

$$\begin{bmatrix} \Delta y \\ \Delta \theta \end{bmatrix} = \begin{bmatrix} k_{11} & k_{12} \\ k_{21} & k_{22} \end{bmatrix}^{-1} \begin{bmatrix} F \\ M \end{bmatrix} = \begin{bmatrix} a_{11} & a_{12} \\ a_{21} & a_{22} \end{bmatrix} \begin{bmatrix} F \\ M \end{bmatrix} \tag{8.18}$$

where

$$a_{11} = \frac{L^3}{3EI}; \; a_{12} = a_{21} = \frac{L^2}{2EI}; \; k_{22} = \frac{L}{EI} \tag{8.19}$$

Equation (8.18) is actually more commonly known. Based on Equation 8.18, if we apply only a downward force F without M, the resulting vertical deformation is $\Delta y = \frac{FL^3}{3EI}$ and the torsional (also called angular) deformation is $\Delta \theta = \frac{FL^2}{2EI}$. Similarly, if we only apply a moment M without the downward force, we have $\Delta y = \frac{ML^2}{2EI}$ and $\Delta \theta = \frac{ML}{EI}$. These results are well known.

Note that the principle of superposition still applies as indicated by Equation (8.18) and Fig. 8.9. This is an important result that will be used in the next section to solve problems that the free-body diagram alone cannot solve.

8.2 Over-Constrained Problems with 2D Springs When FBD is Not Enough

As shown in Fig. 8.10(a), a beam is clamped at both ends and a force F is applied at the center of the beam. The deformation of the beam is shown by the dashed lines. Note that because of the symmetric structure and the clamping ends, the angular deformation at points A, B, and C are all zero, while the vertical deformation at point C is Δy.

If we consider the beam at equilibrium as a rigid body and apply the Free-Body Diagram (FBD) to determine the constraint forces and moments at points A and B, we can draw free-body diagrams as shown in Figs. 8.10(b–c). For the FBD of (c), the boundary is immediately to the right of force F. For the FBD of (b), we have four unknowns. Due to the symmetric structure, we can safely reduce these four unknowns to two, as

$$R_A = R_B = R \text{ and } M_A = M_B = M \tag{8.20}$$

The constraint forces R are readily determined by the force equilibrium in the vertical direction, and we have $R_A = R_B = R = \frac{1}{2}F$. However, when

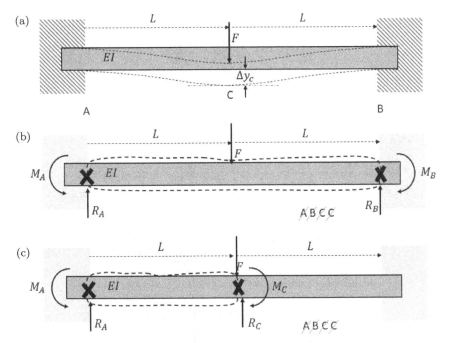

Figure 8.10:   (a) A flexible beam clamped both ends; as a 2D spring; (b–c) FBDs for solving for the constraint forces.

we apply the moment equation with respect to a point on the beam, such as point A, we have

$$\sum M_A = -M + F\,L - \left(\frac{1}{2}F\right)(2L) + M = 0 \qquad (8.21)$$

which is a trivial equation. We cannot determine moment M.

If we use a different FBD as shown in Fig. 8.10(c), we have $R_C = \frac{1}{2}F$ and

$$\sum M_A = -M + F\,L - \left(\frac{1}{2}F\right)(L) + M_C = 0 \qquad (8.22)$$

Equation (8.22) cannot be solved because it has two unknowns, M_C and M.

In order to solve this particular problem analytically, we must consider the beam as a 2D spring and use the information related to the geometric deformations to establish additional equations.

We will redraw the free-body diagram in Fig. 8.10(c) as Fig. 8.11.

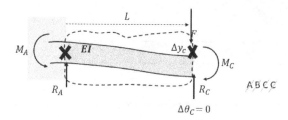

Figure 8.11:   The force analysis considers the doubly clamped beam as a 2D spring.

Per Figs. 8.9 and 8.11, and the definitions of the positive directions, i.e. clockwise rotation as positive and downward as positive, we have the following equation for the 2D spring of Fig. 8.11 as

$$\begin{bmatrix} \Delta y_c \\ \Delta \theta_c \end{bmatrix} = \begin{bmatrix} \Delta y_c \\ 0 \end{bmatrix} = \begin{bmatrix} a_{11} & a_{12} \\ a_{21} & a_{22} \end{bmatrix} \begin{bmatrix} F - R_c \\ M_c \end{bmatrix} \tag{8.23}$$

We now have a new equation related to $\Delta \theta_c$,

$$\Delta \theta_c = 0 = a_{21}(F - R_c) + a_{22}M_c \tag{8.24}$$

Solving Equation (8.24), we have

$$M_c = \frac{-a_{21}(F - R_c)}{a_{22}} = -\frac{1}{4}FL \tag{8.25}$$

The negative value of M_c indicates that M_c should be in the CCW direction.

Putting the value of M_c into Equation (8.22), we can solve for the constraint moments at points A and B as

$$M_A = M_B = M = \frac{1}{4}FL \tag{8.26}$$

The positive values of M_A and M_B indicate that M_A is indeed in the CCW direction and M_B as CW, as assumed in Figure 8.10(b).

Finally, we can solve for the vertical deformation at point C as

$$\Delta y_c = a_{11}(F - R_c) + a_{12}M_c = \left(\frac{L^3}{3EI}\right)(1/2\,F) + \left(\frac{L^2}{2EI}\right)\left(-\frac{1}{4}FL\right) = \frac{FL^3}{24EI} \tag{8.27}$$

As discussed in this example, as well as those in Sections 8.1.4–8.1.6, we often need to use the relationship of elastic deformations to establish additional equations. Although this approach is useful, it is actually difficult to implement when the over-constrained system has a more complex structure.

8.3 The Force Flow Concept

In this section, we introduce a different technique, called the force flow concept, to deal with over-constrained problems. The force flow concept is a powerful tool to get a direct "feel" of the loading condition. Juvinall[2] provided an excellent description of the force flow concept. Several applications were presented to demonstrate how to identify critical sections of a machine, to analyze redundant ductile structures, and to determine stress distributions within axially loaded members. In this section, we will follow the description given in Juvinall's book, with additional illustrations, to demonstrate how to apply the force flow concept for analysis.

As shown in Fig. 8.12(a), if an elastic member is subjected to compressive forces at both ends, the entire element will be subjected to compression. One can imagine that the force at point A is "flowing" to the right, along its line of action, to meet with the force at point B, achieving equilibrium, or vice versa. The passage of the force is identified by a double

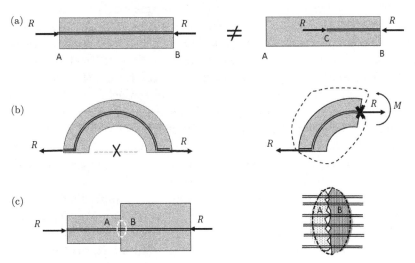

Figure 8.12: Force flow path examples: (a) the force flow through a straight member; (b) the force flow through a curved member and its equivalent FBD; (c) the force flow through two members and their contact points.

---

[2] Juvinall, R.C. (1984). *Fundamentals of Machine Component Design*. Wiley: New York.

line. This double line indicates the section of the beam under compression and along which the force will flow. We can call it the force flow path.

If the force on the left is acting on point C, instead of A, then the flow path only covers the section of the element between points C and B. In other words, only the section of the element between points C and B is under compression. We have two different loading conditions in Fig. 8.12(a). If the rigid body assumption were made, these two cases would be identical because, as discussed in Chapter 1, for a rigid body, a force can move along its line of action without changing the loading condition.

In Fig. 8.12(b), a bow-shaped element is subjected to a pair of tensile forces. The forces cannot flow through the air. They must stay within the element to form a curved force flow path. Note that this force flow path is not the line of action. As shown in Fig. 8.12(b), when the force flows to a new location, it could result in a moment, as indicated by the FBD. When we apply the force flow concept, we do not have to worry about the moment. We only need to identify the section of the element under the influence of the force.

In Fig. 8.12(c), two elements are pressed together by a pair of compressive forces. The force on the left will flow to the right as before until it reaches the interface between the two elements. At the interface, the mating surfaces, in a practical sense, are never perfectly flat. An enlarged view is shown on the right. When looking closely at the mating surfaces, the force can only flow through the contact points at the mating surfaces. At those contact points, if they stay in the elastic region, the material will deform to form a small contact area around each contact point. This contact mechanism is well described by the Hertzian contact stress model. The deformation behavior at the mating surface is highly different from those in the bulk material. Therefore, when we construct a flow path, we must always pay attention to the interfaces. This interface contact deformation will be further explained in the next section.

We illustrate how the force flow concept can be applied for two plates clamped by a bolt and a nut in Fig. 8.13(a). A clamping force, N, is developed after tightening the nut over the bolt. This clamping force is determined by the tightening torque. If there are no external forces, the FBD is shown in Fig. 8.13(a), in which this clamping force is an internal force, not showing up in the FBD as in Fig. 8.13(a). If we isolate the plates and the bolt/nut separately, as shown in Figs. 8.13(b–c), this clamping force would compress the plates, while stretching the bolt. The FBDs of Figs. 8.13(b–c) are not useful because we cannot solve for force N. We will apply the force flow concept for better analysis.

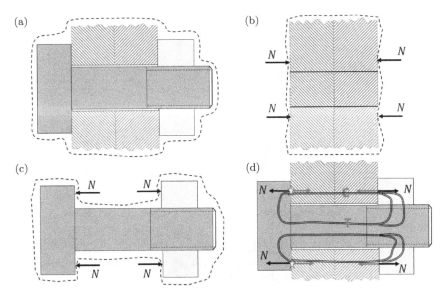

Figure 8.13: (a–c) FBDs for two plates clamped by a bolt and a nut; (d) equivalent force flow diagram.

To construct the force flow of this two-plate assembly shown in Fig. 8.13(d), we can start with the tensile force by the plate to the bolt head, where a star sign is marked. This force will flow into the bolt head and then flow through the bolt shaft to the other end. From there, it will pass through the thread contacts into the nut (two paths are drawn for illustration). The force flow then continues to the interface between the nut and the second plate. From the second plate, it will then pass through the interface between the plates to the plate on the left, thus completing the force flow loop. In this case, a self-contained force flow loop is formed because there are no external forces acting on the assembly. We draw a force loop for either side of the bolt, but typically, we can just draw one loop as long as both sides are identical. A letter "C" is marked at the force flow path where the elements are under compression, while a letter "T" is marked to indicate tensile condition.

An example based on Juvinall is used to show the power of the force flow concept. In Fig. 8.14, two different press designs are shown. If we use FBD to do the analysis, the task becomes highly tedious. However, if we apply the force flow concept, it becomes very clear which design is superior.

First, let us construct a force flow for the design of Fig. 8.14(a). Starting at the interface between press plate B and object A (where the star sign is),

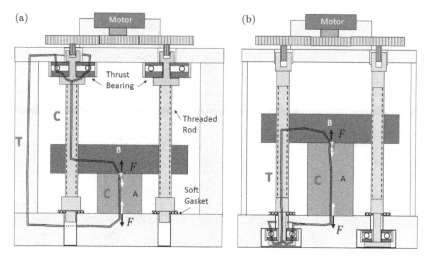

Figure 8.14: Force flow diagrams for design: (a) a press design with thick housing and (b) a different press design with thin housing.

the force first flows into press plate B. Once into plate B, it passes through the thread contacts into the threaded rod. Once in the threaded rod, it could flow up or down, or both. Examining more closely how the threaded rod is held in place, we notice that the force flows into the threaded rod will push the rod upward.

At the lower end of the threaded rod, the rod simply sits on the soft gasket and is inserted into a hole for positioning. There is no constraint at the lower end to resist the upward motion. As a result, if the force flows down along the threaded rod, it has no place to go because it cannot flow into the bottom plate. This means that the force can only flow up along the threaded rod. When the force flows upward along the threaded rod, it will reach the top end, which is attached to the lower plate of a thrust bearing. However, the force can also keep flowing up to the top where the rod is mated to the driving gear. The mating between the threaded rod and driving gear is through the spline coupling.

The spline coupling allows free axial motion but firmly engaged in the torsional direction. The gear can turn the threaded rod while allowing itself to move freely in the vertical direction with respect to the threaded rod. In other words, the vertical movement of the threaded rod is stopped by the thrust bearing only, not by the spline coupling. Now, it becomes clear that the force flows up the threaded rod cannot go into the driving gear due to the spline coupling. It can only flow into the bottom plate of the thrust bearing. From there, the force can flow through the balls into the top plate of the thrust bearing, and finally into the top chassis plate. Once in the top chassis plate, it can only stay in the chassis and flows down through the side chassis plate to the bottom chassis plate and, finally, to the interface between object A and the bottom chassis plate. The force flow loop is completed as shown in Fig. 8.14(a). We then mark the compressive or tensile condition of the flow path. The paths within object A and the threaded rod are compressive, while the one on the side chassis structure is tensile.

Let us construct the force flow for the design of Fig. 8.14(b). We start at the same point, marked by the star sign. The force flows into plate B, and then to the threaded rod. Once in the threaded rod, we observe that the top end of the threaded rod is now floating in the axial direction; therefore, the force can only flow down to the bottom plate of the thrust bearing. Then, it flows through ball contacts to the top plate of the thrust bearing and to the bottom chassis plate. From there, it reaches the interface between the bottom chassis plate and object A, thus completing the loop. We notice that this force flow path does not pass through the side plate of the chassis. In addition, this flow path within the threaded rod is now tensile.

The force flow loops of designs (a) and (b) are very different. It becomes clear that design (b) is superior because (1) the side chassis plate is not under loading; therefore, it can be constructed with thin sheet metal, reducing the overall weight of the press and the material cost; (2) the threaded rod is under tensile force, not in compression. As discussed in Chapter 7 in the truss system, we should avoid putting a long and slender element to compressive forces due to buckling concerns. Design (a) is clearly an inferior design because it costs more while sustaining less press force.

It is so much easier to observe the design differences between the two press designs with the use of the force flow concept. Without it, we will have to disassemble the entire press into many free-body diagrams, and then try to solve several unknown forces in order to detect the above design difference.

8.4 Spring Network Converted from the Force Flow Loop

The force flow passes through many elements of a machine structure, and each element can be considered as a linear spring. If we actually convert each element into a linear spring, the machine structure becomes a spring network.

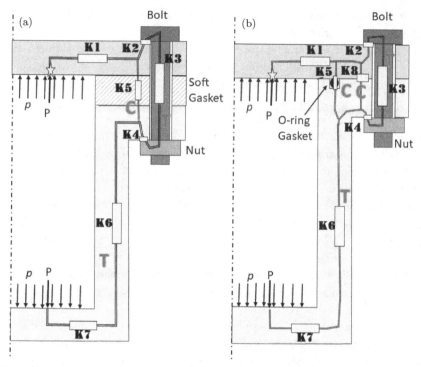

Figure 8.15: (a) Force flow loops of a sealed tank with a soft gasket and (b) Force flow loops of a sealed tank with an O-ring.

We will adopt another excellent example form Juvinall and extend it to illustrate how a force flow loop can be converted into a spring network. After the spring network is constructed, we will discuss how it can be simplified to conduct analysis for critical members of the machine structure, even with over-constrained systems.

As shown in Fig. (8.15), two sealed tanks with high-pressure gas inside are shown. Tank (a) has a gasket between the cover and the tank body to seal the tank, while tank (b) uses an O-ring with the cover directly in contact with the tank body. For tank (b), it is important that the mating surfaces of the cover and the tank body are machined within a proper flatness tolerance to maintain a proper seal.

As illustrated in Fig. 8.15, a force flow loop is plotted for each tank. In addition, we draw a slender rectangle on each force flow path to represent the component that the force flow passes through. With this extension, the new force flow loop can be readily converted into a spring network model.

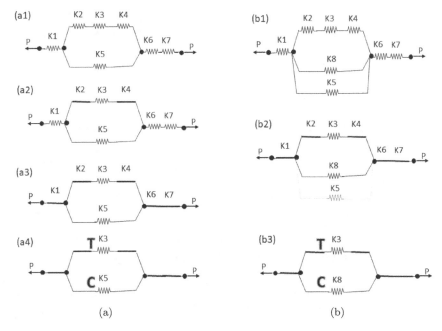

Figure 8.16: Spring network analysis for the tanks in Fig. 8.15.

This is done simply by replacing each component (the slender rectangular) with an equivalent spring. One thing to remember is that an equivalent spring is also needed for the interface between two mating surfaces, as illustrated in Fig. 8.12(c).

The equivalent spring networks are shown in Fig. 8.16. We can trace the force flow to draw the spring network. For example, for the force flow in Fig. 8.15(a), we can start at the top initial point of the force flow path (where a star is marked). We then draw spring k_1. After k_1, the path splits into two paths, one is connected to k_2, then k_3, then, k_4. The other path connects to k_5 and then reconnects with the first path. The joint path then connects to k_6 and k_7, and ends with the other side of the force flow. Redrawing the spring network and making it straight, we have Fig. 8.16(a). In Fig. 8.16(a1), k_2, k_2, and k_3 are in series, and together they are in parallel with k_5.

Because the resulting spring model can be highly complicated when many components are involved, it will be beneficial to conduct preliminary rigidity analysis to simplify the spring network.

There are two simple rules for simplifying the spring network. First, if two springs are connected in series, the combined stiffness is

$$k_{\text{serial}} = \frac{1}{\frac{1}{k_1} + \frac{1}{k_2}} \tag{8.28}$$

where k_1 and k_2 are the stiffness of the springs. If $k_1 \gg k_2$, then the combined stiffness of the two springs in series will be close to the value of the lower stiffness spring; thus, $k_{\text{serial}} \cong k_2$.

Second, if the two springs are connected in parallel, the combined stiffness is

$$k_{\text{parallel}} = k_1 + k_2 \tag{8.29}$$

If $k_1 \gg k_2$, then the combined stiffness of two springs in parallel will be close to the value of the higher stiffness; thus, $k_{\text{parallel}} \cong k_1$.

Based on these two rules, we can simplify a complicated spring network. We first make a judgment on the relative stiffness of each spring. For a rigid spring in series with softer springs, we simply draw a straight line to replace the rigid spring. For a very soft spring in parallel with a much stiffer spring, we simply remove the soft spring from consideration.

First, the two force flow loops of Fig. 8.15 are converted to spring networks as shown in (a1) and (b1) of Fig. 8.16, respectively. We then use the two above rules and the relative stiffness analysis to simplify the spring networks, step by step.

Let us examine the relative stiffness of each spring in spring network (a1) of Fig. 8.16. There are seven springs, k_1 to k_7, to be considered. Springs k_2 and k_4 are due to the Hertzian contact stress between two flat surfaces. The equivalent spring stiffness between two flat metal surfaces is usually very high. Spring k_3 is related to the tension of a slender bolt. Because k_3 is connected in series with k_2 and k_4, the combined stiffness of the springs becomes the value of the softer spring, thus, k_3. As a result, we can replace k_2 and k_4 by rigid lines as shown in (a2) of Fig. 8.16. Springs k_3 and k_4 are in parallel. The combined spring of k_3 and k_4 is in series with k_1, k_6, and k_7. For springs k_1 and k_7, they are related to the bending mode of doubly clamped thick and short structures. We can safely assume that they are of very high stiffness. Spring k_6 is related to the main body of the tank, which is also very stiff. As a result, k_1, k_6, and k_7 can be replaced by rigid lines. We now have a simpler network of (a3). We can add the tensile or compressive signs to (a3), and we have (a4).

Similar simplification procedures can be carried out for spring network (b1) of Fig. 8.16(b). For the tank on the right, spring k_8 is related to the interface between the cover and the tank body, while spring k_5 is related to the O-ring. As these two springs are in parallel, the combined stiffness is close to k_8; thus k_5 can be excluded from the analysis, as shown in (b2). In (b2), springs k_2, k_4, k_1, k_6, and k_7 are also replaced by rigid lines due to their high stiffness values. Finally, we have a simplified spring network of (b3). In both cases (a4) and (b3), we only have two springs in parallel after simplification.

Now let us consider the tank on the left as represented by the spring network (a4) of Fig. 8.16. Before analyzing (a4), we would first consider the case before P is applied. This is the case when the tank cover is mounted and the bolt is tightened to a predetermined clamping load (preloading). This preloading practice causes the bolts to be stretched while the gasket is compressed. Assigning the preloading extension of the bolt δ_{30} and the preloading compression of the gasket δ_{50}, respectively, the preloading clamping force, F_{po}, can be expressed as

$$F_{p0} = k_3\, \delta_{30} = k_5\, \delta_{50} \tag{8.30}$$

After filling in high-pressure gas into the tank, the external force P is now applied as shown in spring network (a4) of Fig. 8.16. The external force, P, causes the bolt to stretch further by an additional extension, δ, while the gasket becomes less compressed by the same amount, δ. In other words, contrary to the common mistake by intuition that all of force P is used to stretch spring k_3, some portion of P is actually used to release the compression of k_5. In other words, P is taken up by both k_3 and k_5. We have

$$P = \Delta F_3 + \Delta F_5 = k_3\delta + k_5\delta \tag{8.31}$$

The forces acting on k_3 and k_5 now become

$$F_3 = k_3(\delta_{30} + \delta) = F_{p0} + \Delta F_3 \tag{8.32}$$

$$F_5 = k_5(\delta_{30} - \delta) = F_{p0} - \Delta F_5 \tag{8.33}$$

From Equations (8.31–8.33), we have

$$\Delta F_3 = Pk_3/(k_3 + k_5) \tag{8.34}$$

$$\Delta F_5 = Pk_5/(k_3 + k_5) \tag{8.35}$$

Note that the $\Delta F_3 + \Delta F_5 = P$.

If a very soft gasket is used, $k_3 \gg k_5$, we have

$$\Delta F_3 \cong P \text{ and } \Delta F_5 \cong 0 \qquad (8.36)$$

In this case, the common intuition is true that the external force P is almost all taken to stretch the bolt, i.e. k_3. If a hard gasket is used, such as those used for the head gaskets in a combustion engine, the stiffness of k_5 could be in the same order of magnitude of k_3. If we assume $k_5 = k_3/3$ (just an estimate), then, based on Equations (8.34–8.35), 75% of P is taken by k_3 (further stretching the bolt), while 25% is taken by k_5 (releasing the compression of the hard gasket). This loading condition will continue until the compression of the gasket is totally released when

$$F_5 = 0 = F_{p0} - P/4 \text{ or } P = 4F_{p0} \qquad (8.37)$$

After that, all external forces P will be taken by the bolt only. Such a condition of total release of the gasket must be avoided because the sealing capability of the gasket will be lost and the high-pressure gas will leak out. In practice, the maximum value of P should be kept well below $4\,F_{p0}$.

The loading and unloading curves of spring network (a4) of Fig. 8.16 are shown in Fig. 8.17. When P is zero, $F_3 = F_5 = F_{p0}$. As P increases, F_3 will increase while F_5 will decrease. The combined value of the increased amount and the decreased amount is equal to P, i.e. $\Delta F_3 + \Delta F_5 = P$.

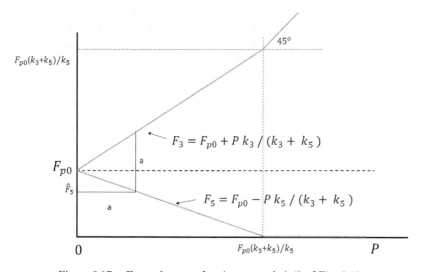

Figure 8.17:   Force changes of spring network (a4) of Fig. 8.16.

When $P = \frac{F_{p0}(k_3+k_5)}{k_5}$, the gasket will be totally unloaded, losing its sealing capability. From that point on, any further increase of P will be taken up by the bolt, making the loading curve increase at a slope of 45 degrees.

For tank (b) in Fig. 8.15, the spring network is similar and the analysis is the same. The difference, however, is that the stiffness of the equivalent spring (K_8) is much higher than that of the gasket $(K_3$ in Fig. 8.15(a). In fact, the stiffness of the equivalent spring of a pair of flat mating surfaces could be higher than the bolt used to tighten them. As a result, a higher preloading clamping force can be achieved when tightening the bolts. The loading diagram for the spring network (b3) of Fig. 8.16 is shown in Fig. 8.18. Note that the value of F_{p0} in Fig. 8.18 is higher than the one in Fig. 8.17. Now, a bigger fraction of external force P is used to unload the compression of the mating surfaces, while a smaller fraction is used for increasing the tension of the bolt. An O-ring is usually used to seal the high-pressure gas. For the gas to leak, the gas will have a substantial pressure drop when it passes through the extremely small gaps between the mating surfaces before it is blocked completely by the O-ring.

If both tanks are used to store gas at the same pressure, the compression force, $\hat{F}_8$, between the two metal surfaces will be higher than the compression force of the gasket, $\hat{F}_5$, as shown in Figs. 8.17 and 8.18. In other words, tank (b) of Fig. 8.15 is capable of storing gas at a higher

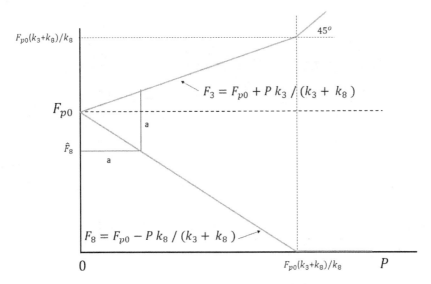

Figure 8.18: Force changes of spring network (b3) of Fig. 8.16.

pressure. The downside is the cost of machining two very flat and smooth surfaces for proper sealing. The O-ring is always used to prevent leaking because surfaces can never be perfectly flat. The tank with a gasket can get away with a surface less flat, but for critical applications, flatness is still required. When the mating surfaces are large, the sealing strategy of tank (b) of Fig. 8.15 becomes more difficult to achieve. This is the case between the engine head and the cylinder block of an internal combustion engine when a hard gasket is used instead of O-rings. Of course, high temperature is also a concern for O-rings. The flatness requirement of a cylinder head is typically $\pm 0.003''$.

It is a common mistake to assume that the bolt is taking up the entire external force without realizing that it also takes some force to release the compression between, for example, the flat mating surfaces. By constructing a force flow loop and converting it to a spring network, the phenomenon becomes fairly easy to understand.

8.5 Applications to Machine Tool Spindle Design

Let us apply the force flow concept and the spring network to a practical machine tool spindle design problem. Figure 8.19 depicts a precision tool-room surface grinder spindle from the Handbook of RHP Bearings.[3] Some additional descriptions are added for better illustration in Fig. 8.19. The installation of this spindle can be described as follows:

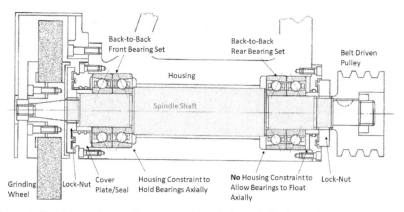

Figure 8.19: A precision tool-room surface grinder spindle from the catalog C826 of RHP Precisions, *High Precision Bearings, Catalog C826*, 1987. Additional descriptions are added for better illustration.

---

[3]RHP Precision. (1987). *High Precision Bearings, Catalog C826*.

Step 1: The front back-to-back bearing set is mounted to the spindle shaft.
Step 2: The spindle shaft is then mounted on to the housing. A cover/seal plate is then mounted. A locknut is then tightened to set a preload.
Step 3: The rear bearing set is mounted to the rear side of the spindle shaft. A rear locknut is tightened to apply a preset preload to the rear bearing set.
Step 4: The remaining components are mounted, such as the grinding wheel and the belt pulley.

As this spindle is designed for cylindrical grinding, the grinding forces are mainly in the radial direction. An axial force is applied if the front face of the grinder is used for grinding. The radial grinding force and its resulting bending moment to the spindle are taken up by both the front and the rear bearings, similar to the door hinge example of Fig. 6.11. The reaction forces in the radial direction of the bearings can be easily determined by assuming the bearings as simple pin-supports. However, for the axial force, it is more complicated. First, we recognize that only the front bearing set can counteract the external axial force because the rear bearing set is designed to be floating axially to accommodate thermal expansions of the shaft. The preloading and the thermally induced preload issues were discussed in Section 8.1.6. We will describe the preloading procedure based on Fig. 8.20 here.

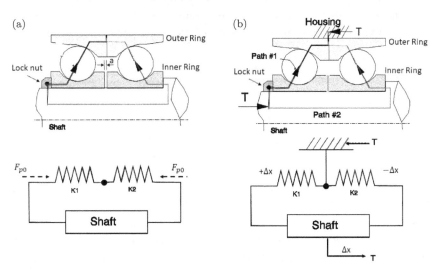

Figure 8.20:   (a) The preloading of a set of back-to-back ball bearings and its equivalent spring network, and (b) the same bearings subjected to an axial loading.

During the preloading stage, as shown in Fig. 8.20(a), there will be no external loading T, and the outer rings of the bearings are not constrained by the housing. When the bearings are mounted onto the shaft, there is a gap, a, between the inner rings. This gap can be adjusted by adding a spacer between the outer rings. The preloading procedure basically is to tighten the locknut so that the gap is closed, i.e. $a = 0$. During preloading, the balls will be compressed, elastically, like a spring, so that the gap can be closed. Once the gap is closed, further tightening will not increase the preload because now the tightening force will transmit directly through the inner rings back to the shaft. If one tries to tighten the locknut forcefully, the threads of the locknut could be stripped. Therefore, it is important that the locknut is 'not over-tightened. Stop once it is tight. At the completion of preloading, the force flow forms a loop, as shown in Fig. 8.20(a). Both bearings are compressed and a preloading force, F_{p0}, is achieved. The shaft is considered as a rigid body because it is much stiffer than the bearing.

Now, consider the axial loading condition illustrated in Fig. 8.20(b). When a thrust force, T, is applied, it shifts the shaft by displacement, Δx. Note that the outer rings are stopped by the housing and the housing is attached to the foundation. This shaft displacement increases the compression in k_1 by Δx (loading), while releasing the compression in k_2 by Δx (unloading). This is similar to the tank examples in Figs. 8.15 and 8.16. However, there is major difference for the equivalent spring of the bearing. The equivalent spring may not be linear, and for bearings, the compression force and the deformation are represented by a nonlinear relationship as

$$F_k = c(\Delta x)^b = (k\,\delta \hat{x})_{\Delta x} \tag{8.38}$$

where c depends on the bearings' type and size, and the value of b is typically 1.5 for ball bearings and 1.11 for roller bearings. The nonlinear relationship can be locally approximated as a linear spring with a linear stiffness, k, if the deformation change, $\delta \hat{x}$, is small, centering around the original deformation, Δx, as indicated by Equation (8.38).

For ball bearings, a typical force vs deformation curve is shown in Fig. 8.21. Once a bearing is preloaded, it reaches an equilibrium point, identified by the dot in Fig. 8.21. If the bearing is further loaded, with a deformation, Δx, the loading point shifts up along the curve to a new point, identified by a star in Fig. 8.21. On the other hand, if the bearing is unloaded, it shifts down to a point, identified by a triangle. The equivalent local linear stiffness constants become k_1 and k_2, respectively, as shown in Fig. 8.21.

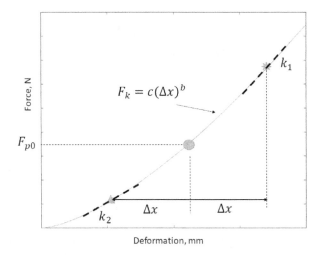

Figure 8.21: The loading curve of a bearing as a nonlinear spring.

RHP Precision, referenced above, based on their empirical testings, found that typically

$$k_1 = 2 \, k_2 \tag{8.39}$$

Based on Equation (8.39), the bearing force changes, using Equations (8.34–8.35), are estimated to be

$$\Delta F_1 \approx T \, k_1 \, / \, (k_1 + k_2) = \frac{2}{3} T \tag{8.40}$$

$$\Delta F_2 \approx T \, k_2 \, / \, (k_1 + k_2) = \frac{1}{3} T \tag{8.41}$$

The above results indicate that two-thirds of thrust force T is taken by bearing k_1 with further compression, while one-third of T is taken by bearing k_2 through unloading. In practical operation of bearings, a certain preloading must be maintained to ensure the "rolling without slipping mode" of operation of the rolling elements; i.e. no ball skidding is allowed to prevent excessive wear and friction. Therefore, it is important that the thrust force T should not exceed 3 times the preload force, F_{p0}, such that bearing k_2 is not totally unloaded.

To increase the axial loading capacity, we can put two bearings in front as shown in Fig. 8.22. In this arrangement, one more spring of k_1 is added in parallel to the original k_1, as shown in Fig. 8.22. As a result, the combined

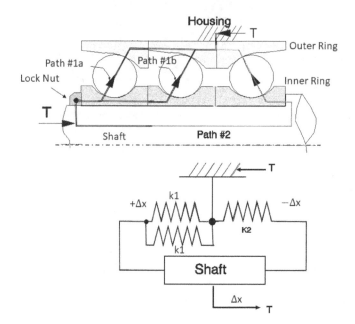

Figure 8.22:   Axial loading analysis for a triple bearing set.

stiffness is $2k_1$. The resulting force changes become

$$\Delta F_1 \approx T \left(2k_1/(2k_1 + k_2)\right) = \frac{4}{5}T \tag{8.42}$$

$$\Delta F_2 \approx T \; k_2/(2k_1 + k_2) = \frac{1}{5}T \tag{8.43}$$

Now the spindle can handle an axial load up to five times the preloading force.

8.6 Concluding Remarks

In this chapter, we discussed the situation when the FBD analysis is either too difficult or not effective in solving problems. This happens often in practical machines when the structures are complex and over-constrained. We also released the rigid body assumption and introduced the linear spring for representing each member of a machine. Most importantly, a new technique, called the force flow analysis, was introduced to allow for visual inspection of the loading conditions of a complex structure. A force flow loop can be directly translated into a spring network, which can then be simplified for effective force analysis for complex structures. Practical examples such as pressure vessels and spindle systems are used for illustration.

Chapter 9

Are all Equilibrium Conditions Created Equal?

In Chapter 8 when we discussed the force flow concept with the example of two presses (Figure 8.14, Chapter 8), we pointed out that we should avoid applying compressive forces to a long and slender element due to the concern of buckling. It would be less of a concern for the same amount of the force when it is applied as a tensile force. In both cases, we can reach static equilibrium. This brings up the question: Are all equilibrium conditions created equal?

9.1 Static Equilibrium versus Steady State

First, let us revisit the definition of the static equilibrium. A system is considered to be at the static equilibrium when the loading conditions satisfy the following two equations:

$$\sum_i \vec{F}_i = 0 \tag{9.1}$$

$$\sum_i \vec{M}_i = 0 \tag{9.2}$$

These two equations do not specify how the equilibrium is reached or how well the equilibrium can be maintained, which will be the main topics discussed in this chapter.

Often the static equilibrium condition is confused with the steady state condition. A steady state condition is defined as the motion of the system that does not change with respect to time or demonstrates a repeated

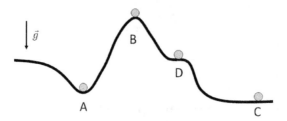

Figure 9.1:   Different types of stability.

pattern with respect to time. A static equilibrium condition is a steady state condition, but a steady state condition is not necessarily a static equilibrium condition. For example, a car moving around a perfect circle at a constant speed is at a steady state condition, but it is not at a static equilibrium condition.

9.2 Stability — How Well can the Static Equilibrium be Maintained

Let us consider a ball resting on different surfaces to illustrate different equilibrium conditions and their stabilities. In Fig. 9.1, the stability in case (A) is a stable equilibrium, case (B) is an unstable equilibrium, case (C) is a neutral equilibrium, while case (D) is a limited neutral equilibrium. In case (A), if you shift the ball away from its equilibrium point, it will return to its equilibrium position (assuming we have a little friction so it will not oscillate around the equilibrium point forever). Therefore, it is considered a stable equilibrium. In case (B), if you shift the ball off the equilibrium point, it will slide down and away; thus, it is an unstable equilibrium. In case (C), if you shift the ball to a new position, it stabilizes at the new position, not returning to the original equilibrium position. Case (C) is different from case (A) and is denoted as a neutrally stable equilibrium. In case (D), it is a combination of case (C) and case (B). If the ball is shifted too far to the right, it will slide off. We can call it conditionally stable equilibrium.

9.2.1 How geometry affects stability

Let us consider how the geometry of an object on a neutrally stable surface affects its stability. As shown in Fig. 9.2(a), a slender rod can achieve an equilibrium in a vertical position, but it is an unstable equilibrium,

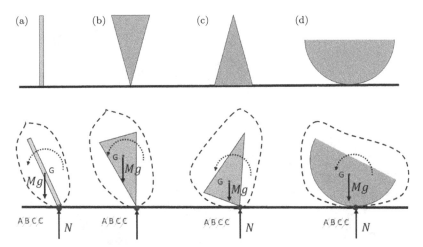

Figure 9.2: The geometry of the object can affect the stability of its equilibrium position: (a) an inverted pendulum and its FBD; (b) an inverted triangle and its FBD; (c) a triangle and its FBD; (d) a semi-disk and its FBD.

as indicated by the corresponding free-body diagram analysis. A moment due to gravity and the contact force from the ground will topple the rod.

Similarly, Fig. 9.2(b), an inverted triangle, theoretically, can reach an equilibrium, but it is unstable. It will topple with a little disturbance. In both cases, the contact to the ground is a point contact. However, in Fig. 9.2(c), the same triangle is now in a stable equilibrium. The moment can return the triangle back to its equilibrium position if it is not tilted too much. In Fig. 9.2(c), the contact of the triangle with the ground is a line or surface contact until it is tilted. In Fig. 9.2(d), even though the contact of the semi-circle (or semi-sphere) is a point contact, the geometry makes it stable. The semi-circle will be returned to its original stable position by a moment when it tilts. The semi-circle can oscillate a few times and, through friction, it will be at static equilibrium again. An example of Fig. 9.2(d) is a roly-poly toy. There are many practical examples of the roly-poly stability, as shown in Fig. 9.3.

In Fig. 9.3(a), the stability of a boat is ensured if the buoyance force and the gravitational force form a moment to return the boat to its equilibrium position. Figure 9.3(b) depicts the famous roly-poly toy. Other applications include a rocking chair (Fig. 9.3(c)), a safe iron (Fig. 9.3(d)), and a piece of drinking glass that will not topple (Fig. 9.3(e)).

The case shown in Fig. 9.3(f) is related to the steering of a tire. If the car is making a right turn, as shown, a resulting steering force, R, is

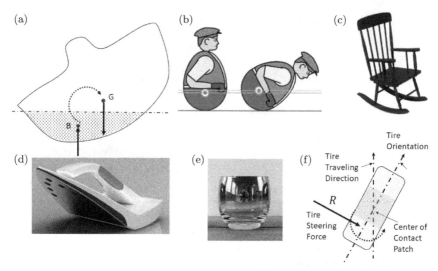

Figure 9.3:   Different applications of roly-poly stability: (a) boat; (b) rolly-polly toy; (c) rocking chir; (d) burn-proof iron; (e) non-toppling drinking glass; (f) returning action of a tire.
Source: (b) https://upload.wikimedia.org/wikipedia/commons/thumb/2/20/Poli_Gus_N_rocked.svg/320px-Poli_Gus_N_rocked.svg.png; (c) https://autumnsunshineandgabrie lleangel.files.wordpress.com/2011/10/r-chair.jpg?w=220; (d) http://2.bp.blogspot.com/ -2M7QaLD4rfY/UHrtgCN9Wgl/AAAAAAABPal/Qsj3lKz2kyDE/s1600/3.jpg; (e) http ://c1.staticflickr.com/9/8022/7686917550_e444ffb67c.jpg.

formed by the ground to the tire. This force makes the car turn to the right. However, if the driver let the steering wheel go, the tire will have a tendency to rotate back (to the left). This is due to the steering force that is located behind the center of the contact patch so that it forms a return moment to turn the tire back, at least partially.

9.2.2 How force directions affect stability

So far, all the examples of stability shown in Figs. 9.1–9.3 are subjected to compressive forces at equilibrium. Their equilibrium positions could be stable or unstable depending on their geometry and contact conditions. However, if the equilibrium is maintained through a tensile force pair, the equilibrium tends to be stable. This concept is shown in Fig. 9.4. The slender rod and the inverted triangle in their vertical positions on the ground are unstable. However, if we hang the rod with a wire, it is now stable at a vertical position. At equilibrium, the forces are a pair of tensile forces, as shown in Fig. 9.4(b). Similarly, the inverted triangle can be hung in different

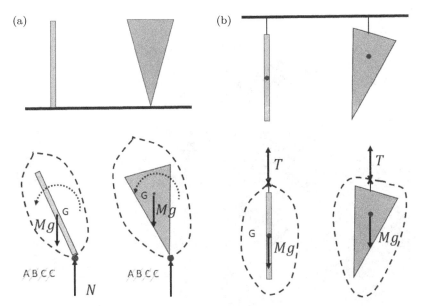

Figure 9.4: (a) Equilibrium achieved by compressive force pair tends to be unstable and (b) equilibrium achieved by tensile force pair tends to be stable.

vertical positions as stable equilibrium. Naturally, both tensile forces will pass through the mass center.

In our daily life, there are many stability cases that have choices of either tensile or compressive forces, such as pushing a shopping cart versus pulling it. We will have to control the shopping cart direction when pushing, but the shopping cart will just follow you when you pull it behind you. It becomes more obvious if you have to push three carts versus pulling. Now, if you have to handle 20 shopping carts as a worker for a supermarket, it is very hard to push them. They are pulled with the help of a machine, as shown in Fig. 9.5. If not locked rigidly, the shopping carts will not be able to stay in a straight line if you push them, which is a phenomenon similar to buckling. Similarly, a horse is usually arranged to pull a wagon, not push it. For trains, a locomotive (or train engine) typically pulls the multiple units of railcars — it does not push them. Some train systems might have a locomotive in front to pull and the other at the end to push, but the case of pushing only without pulling is very rare. Figure 9.6 illustrates how logs are pulled in a river by a boat.

As stated, pulling has inherent stability, while pushing has the advantage of seeing what you are pushing without turning your head back or

Figure 9.5:   Pushing or pulling a long line of shopping carts with a motorized tool.
Source: https://upload.wikimedia.org/wikipedia/commons/thumb/3/33/Shopping_cart
_shepherd_for_Target_jeh.jpg/1280px-Shopping_cart_shepherd_for_Target_jeh.jpg.

Figure 9.6:   Logs are pulled on a river.
Source: http://c1.staticflickr.com/1/55/162774655_c0db49beef.jpg.

Figure 9.7: Barges are pushed in Mississippi river.
Source: http://upload.wikimedia.org/wikipedia/commons/8/8b/Mississippi_river_barg es_after_passing_under_the_Poplar_St_Bridge,_St_Louis,_MO.jpg.

using a rear-view mirror for monitoring. Interestingly, barges are pushed instead of being pulled. Figure 9.7 illustrates how barges are pushed by a boat on the Mississippi river. However, these barges are rigidly tied together and would not twist (or buckle).

The advantage of pushing is the provision of better view while steering. A big cargo ship is pushed from the stern of the ship. The first airplane by the Wright brothers had the propeller in front, that is, pulling. However, rockets have the propulsion system at the rear, pushing the payload up. Rocket guidance stability is a notoriously difficult problem. For cars, we have front-wheel drive and rear-wheel drive versions. As a car body is not too long and rather rigid, pushing or pulling it does not create that much of a difference. However, when a car is braking, the stability related to pushing and pulling becomes significant.

As shown in Fig. 9.8(a), if a car is moving to the right and a force is applied at the rear to stop the car, it would be similar to the pulling case with a better steering stability. As a result, the car has a better chance of staying straight. In Fig. 9.8(b), if the force is applied in front to stop the car, this is equivalent to the pushing case. If the force is not properly aligned, the car can spin. In practice, as shown in Fig. 9.8(c), braking forces are applied by the ground to both the front and the rear tires. Due to the weight shift to the front, the braking force in front is higher (i.e. bigger rotors and brake pads in the front brake). As a result, the actual braking

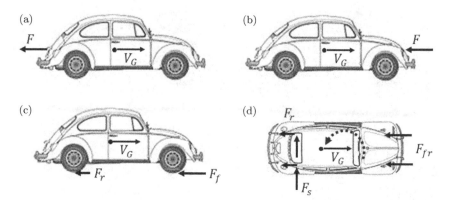

Figure 9.8: The braking force distribution could affect the steering stability during braking: (a) pulling from behind; (b) pushing from front; (c) higher braking force in front; (d) side force needed to keep the car straight during braking.
Source: https://upload.wikimedia.org/wikipedia/commons/thumb/2/24/Quatro_vistas_do_fusca.jpg/250px-Quatro_vistas_do_fusca.jpg.

is more similar to Fig. 9.8(b), and there is a concern of the car spinning during braking. To keep the car from spinning, as shown in Fig. 9.8(d), a lateral force, F_s, at the rear tires is developed to keep the car straight. The combined force of F_r and F_s cannot exceed the static friction of the rear tire. When it is raining, the traction, or achievable frictional force of tires, is greatly reduced due to hydroplaning. A car is more likely to spin due to the reduced traction. Therefore, if one can only afford two new tires, these two new tires should be installed at the rear, no matter if the car is a front-wheel or rear-wheel drive.

The lateral force of the tire is considered in Problem 11.30 of Chapter 11.

9.3 Mathematical Criteria for Stability

In the previous sections, we introduced different equilibrium conditions and their corresponding stabilities. However, all discussions on stability are based on qualitative descriptions or intuitions. In this section, we will introduce a mathematical definition of equilibrium and a criterion of stability.

9.3.1 Work and energy

We need to introduce the concepts related to energy and work for the mathematical description of stability. First, what is work? Or more precisely, what is the work done by a force?

Mathematically, we define an infinitesimal work, δW, done by a force, $\vec{F}$, as

$$\delta W = \vec{F} \cdot d\vec{r} = F dr \cos \theta \qquad (9.3)$$

where $d\vec{r}$ is an infinitesimal displacement and θ is the angle between $\vec{F}$ and $d\vec{r}$. The unit of work is the same as energy. For the SI system, 1 joule (J) of energy is equal to $1\,\mathrm{Nm}$, or a force of one newton (N), acts on an object that moves in the same direction of the force through a distance of one meter (m). We will not go into the English units on energy and work. Basically, work and energy are exchangeable terms, but we cannot use one instead of the other in some sentences, such as "An energy is done by a force..."

If $\theta = 90°$, then δW is zero. Of course, δW can be zero if either $\vec{F}$ or $d\vec{r}$ is zero.

For example, when a satellite is orbiting around the Earth in a circular path, the force applied by the Earth is always perpendicular to the movement of the satellite; as a result, the Earth does not do any work on this satellite. Similarly, if you tie a rock with a string and spin it around, the force of the string does no work on the rock.

The definition in Equation (9.3) appears to be simple, but conceptually, there is a problem. For all the discussions we have about forces, we define a force as a vector, which has a magnitude, a direction, and a point of application. However, we never stated that the force has a motion. As a result, the infinitesimal displacement, $d\vec{r}$, is not about the little displacement of the force, $\vec{F}$, but the motion of the object at the point of application. Basically, there is no motion concept of the force.

Let us explain this with the example in Fig. 9.9.

In Fig. 9.9, a rigid body, object A, has an infinitesimal displacement, $d\vec{r}$, at this instant of time along a horizontal plane. The free-body diagram

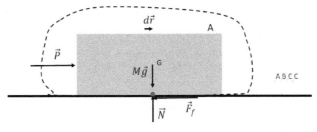

Figure 9.9:   Work done by different forces on a rigid body with a small displacement, $d\vec{r}$.

analysis shows that there are four forces acting on the object, a horizontal external force, $\vec{P}$, a body force, $M\vec{g}$, a frictional force, $\vec{F}_f$, and a normal force, $\vec{N}$, with corresponding points of applications.

As object A is moving with $d\vec{r}$, we only state that the forces in Fig. 9.9 will act on the object as shown, without commenting on how these forces will move together with the object. Based on Fig. 9.9, there is no work done by $M\vec{g}$ and $\vec{N}$ because they are perpendicular to $d\vec{r}$. The works done by $\vec{P}$ and $\vec{F}_f$ are

$$\delta W_1 = \vec{P} \cdot d\vec{r} = P dr \quad \text{and} \quad \delta W_2 = \vec{F}_f \cdot d\vec{r} = -F_f dr \qquad (9.4)$$

Force $\vec{P}$ apparently needs to move along with block A in order to have an effect on it. However, we do not consider the movement of the force. Similarly, frictional force $\vec{F}_f$ does not stay in one place. We only consider the motion of the object.

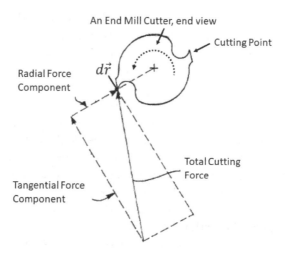

Figure 9.10: The machining force components of end milling and the work done by them.[1]

[1] Jeppsson, J. (1989). "Method for indicating end mill wear". US Patent 4802095, Jan. 31, 1989.

Tu, J.F. and Corless, M. (2014). "Review of sensor-based approach to reliable high speed machining at boeing — A tribute to Jan Jeppsson, a trailblazer for HSM implementation". *High Speed Machining.* **1**(1), ISSN (Online) 2299-3975, DOI: 10.2478/hsm-2014-0001, August 2014.

Jeppsson, J. (1999). "Sensor based adaptive control and prediction software — Keys to reliable HSM," SME APEX '99 Machining and Metalworking Conference, Detroit, Sept. 15.

In machining, such as milling, the total cutting force at the cutting point can be represented by its radial component and its tangential component as shown in Fig. 9.10. The cutting tool, called end mill, in Fig. 9.10 is rotating, and the cutting point has a $d\vec{r}$ movement. As a result, the radial cutting force does not do any work during machining, only the tangential force does. The energy needed to remove materials during machining is supplied by the spindle motor, and it is only related to the tangential force. One would think that the radial force is not important during machining. It turns out that the radial force is very sensitive to tool wear, which affects the efficiency of the machining. A patent was awarded to Boeing to detect the tool wear based on the ratio of the radial force to the tangential force (see footnote 1).

9.3.2 Conservative forces

Often, it is more important to know the work done over a finite displacement of the object, not just infinitesimal work as defined in Equation (9.3). We can simply integrate Equation (9.3) as

$$\Delta W = \int \delta W = \int_{r_1}^{r_2} \vec{F} \cdot d\vec{r} \tag{9.5}$$

If we know both $\vec{F}$ and $d\vec{r}$ as functions of time, we can integrate Equation (9.5) to find ΔW. However, in general, such information is not available, unless we install sensors to measure both. The condition for which we can carry out the integration is when $\vec{F}$ is a function of $\vec{r}$. What does it mean that $\vec{F}$ is a function of $\vec{r}$? This means that the size and direction of the force is only related to the position vector of the object it is acting on. If so, we can carry out the integration of Equation (9.5) and the work done, $\Delta W(r_1, r_2)$, is only a function of the initial position (r_1) and the final position (r_2) of the object,

$$\Delta W(r_1, r_2) = \int_{r_1}^{r_2} \vec{F}(\vec{r}) \cdot d\vec{r} \tag{9.6}$$

The forces with this property are very "nice" because we can readily determine the work done by them. These forces have a special name: Conservative Forces. There are several specific examples of conservative forces important to engineering analysis. The first example is the gravitational force. For an object near the Earth's surface, the gravitational pull is $\vec{F} = M\vec{g}$, which is a constant both in terms of direction and size. As a

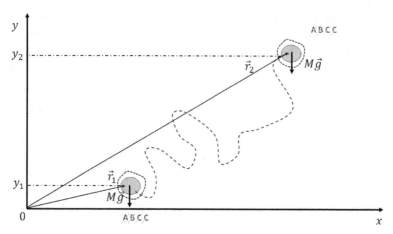

Figure 9.11: An object moves through an arbitrary path from a new position $\vec{r}_1$ to a new position, $\vec{r}_2$. The work done is independent of the path it has traveled.

constant, it is indeed a function of position because we can easily carry out the integration as in Equation (9.6).

As shown in Fig. 9.11, an object of mass M moves from position $\vec{r}_1$ to position $\vec{r}_2$. The work done by the gravitational force is

$$\Delta W(r_1, r_2) = \int_{r_1}^{r_2} \vec{F}(\vec{r}) \cdot d\vec{r} = \int_{y_1}^{y_2} M g \cos(180^\circ) dy$$

$$= \int_{y_1}^{y_2} -Mg \, dy = Mgy_1 - Mgy_2 \qquad (9.7)$$

which is only a function of the initial and the final positions, independent from the path it travels from $\vec{r}_1$ to $\vec{r}_2$. From Equation (9.7), we can also define a new energy term, denoted as gravitational potential energy, $U_g = Mgy$. The work done by the gravitational force is only the difference between the gravitational potential energy,

$$\Delta W(r_1, r_2) = Mgy_1 - Mgy_2 = U_{g1} - U_{g2} = -\Delta U_g \qquad (9.8)$$

From Equation (9.8), it becomes quite straightforward for us to determine the work done by the gravitational force. If an object moves from a lower position to a higher position, the work is negative, while the work is positive if it moves from a high position to a lower position. In other words, a positive work done by the gravitational force will reduce the gravitational potential energy, i.e. $\Delta W_g = -\Delta U_g$. Finally, the datum to determine the vertical position is not unique. We can define any vertical position as the datum, i.e. $y = 0$. The resulting work done by the gravitational force will be the same. However, we must use the same datum throughout the calculation.

We do not invoke the concept of kinetic energy in the discussion above, which is a topic in Dynamics.

Let us also answer a simple question. Because the gravitational force near the Earth is conservative and constant, is every constant force a conservative force? For example, if a block is moving on a flat surface with a constant friction force, is this constant friction force a conservative force? The answer is NO because while the magnitude could be constant for this friction force, its direction will change when the block is moving toward a different direction. The magnitude can also change if it moves to a different surface or there is an external force acting on it.

Another example of the conservative forces is related to the spring, linear or nonlinear. We will discuss only the linear spring case. As shown in Fig. 9.12, an ideal linear spring with an original length of x_0 has been stretched by x. At that point, the spring force on the free end is $\vec{F}_k$. If the spring is further stretched by an infinitesimal amount, dx, the spring force would be essentially the same, but there is an infinitesimal work done by $\vec{F}_k$ as

$$\delta W = \vec{F}_k \cdot d\vec{x} = -F_k dx \tag{9.9}$$

Here, we must adhere to the definition of the work and define δW based on the FBD in Fig. 9.12.

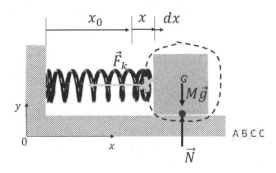

Figure 9.12:   A linear spring is stretched from its original length.

If the spring is stretched to x_2, then the work done by the spring force is

$$\Delta W(x_1, x_2) = \int_{x_1}^{x_2} -F_k \, dx$$

$$= \int_{x_1}^{x_2} -kx \, dx = \frac{1}{2}kx_1^2 - \frac{1}{2}kx_2^2 = U_{k1} - U_{k2} = -\Delta U_k \quad (9.10)$$

where $F_k = kx$ and $U_k = \frac{1}{2}kx^2$.

Here we have the definition of a new potential energy, $U_k = \frac{1}{2}kx^2$. The work done by the spring to the block only depends on the potential energies at its initial stretch and the final stretch, similar to the gravitational force case. Also similar to the gravitational potential energy, a positive work by the spring force is to reduce the spring potential energy, i.e. $\Delta W_k = -\Delta U_k$. Note that the spring potential energy is positive either if the spring is compressed or stretched. The datum for calculating the spring potential energy is defined when there is no deformation in the spring. We cannot arbitrarily define a datum for the spring deformation. U_k is also called elastic potential energy because the spring is assumed to be elastic and does not suffer permanent deformation (plastic deformation).

Now, let us combine these two conservative forces in an example. As shown in Fig. 9.13, object A is tied to a linear spring, with a spring constant k, on a frictionless slope and reaches an equilibrium with the spring compressed by z_1 from its original length z_0. We can easily find out z_1 by the equilibrium force equations, $z_1 = \frac{mg \sin \theta}{k}$.

Now, let us examine the potential energy condition at the equilibrium for the system of object A and the spring. Three forces are involved in Fig. 9.13. The normal force $\vec{N}$ is not a conservative force, but it does not do any work in this case because it is always perpendicular to the motion direction of the block. With respect to datum $O - x - y$, the total potential energy is

$$U_{\text{total}} = U_{k1} + U_{g1} = \frac{1}{2}kz_1^2 + (z_0 - z_1)mg \sin \theta \quad (9.11)$$

Now, the question is "Does the total potential energy have any specific properties at equilibrium?" The answer is YES! At equilibrium, the total potential energy should be either at minimum or at maximum. Let us just accept this conclusion for now and check if this condition leads to the same conclusion we had for the amount of spring compression.

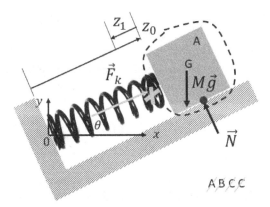

Figure 9.13: Potential analysis involves both gravity and spring.

If the total potential energy is either at maximum or minimum, its derivative is zero,

$$dU_{\text{total}} = kz_1 dz_1 - mg \sin \theta \ dz_1 = 0 \tag{9.12}$$

Solving the above equation, we have $z_1 = \frac{mg \sin \theta}{k}$, same as the result based on the force equilibrium.

Are we just lucky? In fact, Equation (9.12) and the force equilibrium equation are the same for the one-particle system of Fig. 9.13. At equilibrium, we have

$$\sum \vec{F} = 0 \tag{9.13}$$

If the only forces involved which could do work are conservative forces, we can multiply Equation (9.13) with $d\vec{r} = d\vec{z}_1$ (assuming that object A is moving up by an infinitesimal amount)

$$\sum \vec{F} \cdot d\vec{r} = \delta W_k + \delta W_g = kz_1 dz_1 - mg \sin \theta \ dz_1 = 0 \tag{9.14}$$

Equation (9.14) is zero because $\sum \vec{F} = 0$. This proves that Equations (9.12) and (9.13) are actually identical. Equation (9.14) is a new method to determine the equilibrium condition, which is denoted as Method of Virtual Work. We will extend the method of virtual work from one particle to a rigid body and a combination of many rigid bodies in the next section.

Note that in order to use the total potential energy (Equations (9.11) and (9.12)) to predict the equilibrium condition, only conservative forces should be involved. For the example of Fig. 9.13, if there is friction between block A and the slope surface, then the equilibrium position is no longer uniquely defined, which means that block A could reach equilibrium at a different compression of the spring when there is a friction. However, we can still use Equation (9.14) by including non-conservative forces. The method of virtual work basically is a method to assume a virtual displacement ($d\vec{r}$ or $\delta\vec{r}$) from the equilibrium position, then all virtual works done by all forces (conservative or non-conservative) should be zero, or

$$\sum \delta W = \sum \delta W_{,\text{non}} - \sum dU = 0 \qquad (9.15)$$

where $\sum \delta W_{,\text{non}}$ is the work done by non-conservative forces and $\sum \delta W_{,\text{con}} = -\sum dU$ is the work done by conservative forces, which is equal to the negative change of the corresponding potential energy as described by Equations (9.8) and (9.10).

Let us consider the same system of Fig. 9.13 to show how the method of virtual work can be applied when non-conservative forces are involved. A non-conservative force $\vec{P}$ is now applied to move object A from its original position to a new equilibrium position. At the new equilibrium position, the force equilibrium is achieved with both the effects of $\vec{P}$ and frictional force $\vec{F}_f$ (Fig. 9.14). Let us consider a simpler friction condition by assuming that $F_f = \mu_d N$. Using the equilibrium force equation, we can easily find out that $z_2 = \frac{(\mu_d \cos\theta + \sin\theta)\, mg - P}{k}$. Let us try the method of virtual work for the same problem. Assuming a virtual displacement, dz, moving up from the new equilibrium position, the sum of all corresponding virtual works becomes

$$\sum \delta W = \delta W_k + \delta W_g + \delta W_f + \delta W_N + \delta W_P = 0 \qquad (9.16)$$

where

$$\delta W_k = \vec{F}_k \cdot d\vec{z} = k z_2\, dz = -dU_k$$

$$\delta W_g = m\vec{g} \cdot d\vec{z} = -mg \sin\theta\, dz = -dU_g$$

$$\delta W_f = \vec{F}_f \cdot d\vec{z} = -\mu_d N\, dz$$

$$\delta W_N = \vec{N} \cdot d\vec{z} = 0$$

$$\delta W_P = \vec{P} \cdot d\vec{z} = P\, dz$$

From Equation (9.16), we easily obtain the same result for z_2.

Before we move on to the method of virtual work for rigid bodies, let us comment on the difference between the force equilibrium

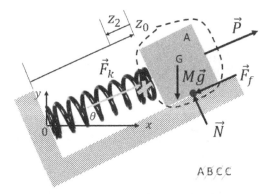

Figure 9.14: New equilibrium position when non-conservative forces are involved.

equation (Equation (9.13)) and the virtual work equation (Equation (9.15)). Equation (9.13) is a vector equation, which means that it actually contains two equations for a 2D problem and three equations for a 3D problem. However, Equation (9.15) is a scalar equation and it is only one equation, be it a 2D or 3D problem. Therefore, Equation (9.15) is easier to use but can only be used to solve for one unknown.

9.4 Method of Virtual Work for Rigid Bodies

In the last section, we showed the method of virtual work for one particle. Let us extend it to a rigid body and then to a system with multiple rigid bodies. As we discussed before, a rigid body is an object that never deforms. As a result, the distance between any two points of the rigid body must stay the same under any loading conditions. If the rigid body is undergoing any motion, in order to keep the rigid body condition, these two points must either have the same movement, or one must move around the other point as a perfect circular motion. The angular velocity of this circular motion must be the same for any point. You can identify three points on the rigid body that form a triangle. The shape and size of the triangle must stay the same. The consequence of the rigid body condition is that the motion of any point on a rigid body can be represented by the motion of a point and the rotation of the rigid body. As shown in Fig. 9.15, the velocity of any point of the rigid body can be represented as

$$\vec{V}_{i/O} = \vec{V}_{G/O} + \vec{\omega} \times \vec{r}_{i/G} \qquad (9.17)$$

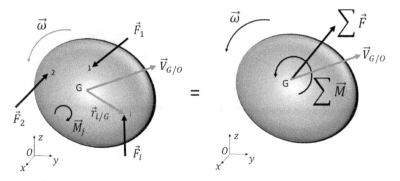

Figure 9.15:   The loading condition and motion of a rigid body.

where $\vec{V}_{i/O}$ is the velocity of point i with respect to a reference frame $O - x - y - z$, $\vec{V}_{G/O}$ is the velocity of point G, $\vec{\omega}$ is the rotational velocity of the rigid body, which is a universal property of the rigid body, and $\vec{r}_{i/G}$ is the relative positive vector of point i to point G. Equation (9.17) is one of the most important equations of rigid body kinematics. We introduce it here for the sake of deriving the virtual work principle for rigid bodies. Also note that $\vec{V}_{i/O} = \frac{d\vec{r}_{i/O}}{dt}$ or $d\vec{r}_{i/O} = \vec{V}_{i/O}dt$, and $d\vec{\theta}_{i/O} = \vec{\omega}_{i/O}dt = \vec{\omega}\,dt = d\vec{\theta}$.

For the rigid body in Fig. 9.15, there are several forces and moments acting at different points. All these forces and moments can be moved to point G (mass center, for example, but it does not have to). At equilibrium, we have both force and moment equilibrium

$$\sum \vec{F} = \sum \vec{F}_i = 0 \tag{9.18}$$

$$\sum \vec{M} = \sum (\vec{F}_i \times \vec{r}_{i/G}) + \sum \vec{M}_j = 0 \tag{9.19}$$

The total virtual work done by all the forces ($\vec{F}_i$) and moments ($\vec{M}_j$) for their respective infinitesimal linear displacements, $d\vec{r}_{i/O}$, and angular displacements, $d\vec{\theta}_{i/O} = d\vec{\theta}$, is

$$\sum \delta W = \sum \delta W_F + \sum \delta W_M = \sum \vec{F}_i \cdot d\vec{r}_{i/O} + \sum \vec{M}_j \cdot d\vec{\theta}_{j/O} \tag{9.20}$$

Introducing Equation (9.17) into Equation (9.20), we have

$$\sum \delta W = \left(\sum \vec{F}_i \cdot \vec{V}_{i/O} \right) dt + \left(\sum \vec{M}_i \right) \cdot \vec{\omega}\,dt$$

$$= \left(\sum \vec{F}_i \cdot (\vec{V}_{G/O} + \vec{\omega} \times \vec{r}_{i/G}) \right) dt + \left(\sum \vec{M}_j \right) \cdot \vec{\omega}\,dt$$

$$= \left(\sum \vec{F}_i \right) \cdot \vec{V}_{G/O} dt + \left(\sum \vec{F}_i \cdot (\vec{\omega} \times \vec{r}_{i/G}) \right) dt + \left(\sum \vec{M}_j \right) \cdot \vec{\omega} \, dt$$

$$= \left(\sum \vec{F}_i \right) \cdot d\vec{r}_{G/O} + \left(\sum \vec{F}_i \times \vec{r}_{i/G} + \sum \vec{M}_j \right) \cdot d\vec{\theta} \qquad (9.21)$$

In the above derivation, we invoke $\vec{F}_i \cdot (\vec{\omega} \times \vec{r}_{i/G}) = (\vec{F}_i \times \vec{r}_{i/G}) \cdot \vec{\omega}$. This equality can be proved easily by carrying out the cross and dot products on each side.

Introducing Equations (9.18) and (9.19) into Equation (9.21), we have $\sum \delta W = 0$, and therefore Equation (9.20) becomes

$$\sum \delta W = \sum \vec{F}_i \cdot d\vec{r}_{i/O} + \left(\sum \vec{M}_j \right) \cdot d\vec{\theta} = 0 \qquad (9.22)$$

Equation (9.22) is the virtual principle for a rigid body, extended from that for one particle.

Let us extend the virtual work principle to a system of multiple rigid bodies. Consider a system of two rigid bodies connected at point C. We will require that this connection must render the displacements at point C on both bodies to be the same. For example, if these two bodies are pinned together, then the linear displacements will be the same. However, as the pin allows for rotation, then the rotational displacements are not the same. In this case, then we have to have a frictionless pin. Without this requirement, the principle of virtual work will need to include this frictional torque at the connection. We will not consider this condition here. As shown in Fig. 9.16, there are many forces acting on rigid body A, $\vec{F}_{A1} \ldots \vec{F}_{Ai}$, and on rigid body B, $\vec{F}_{B1} \ldots \vec{F}_{Bj}$. If we draw a free body diagram separately for each of them, we have the constraint force $\vec{f}_C$, which will be in the opposite directions for A and B. However, the displacements at point C are the same as $d\vec{r}_C$.

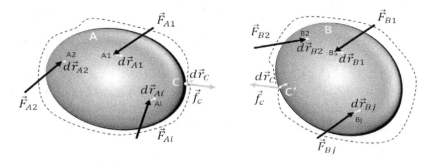

Figure 9.16: A system of two rigid bodies connected at point C.

If the system is at equilibrium, we can apply the principle of virtual work separately to A and B. We have

$$\left(\sum \delta W\right)_A = \sum \vec{F}_{Ai} \cdot d\vec{r}_{Ai/O} + \sum \vec{f}_c \cdot d\vec{r}_c = 0 \qquad (9.23)$$

$$\left(\sum \delta W\right)_B = \sum \vec{F}_{Bj} \cdot d\vec{r}_{Bj/O} - \sum \vec{f}_c \cdot d\vec{r}_c = 0 \qquad (9.24)$$

Adding Equations (9.23) and (9.24), we have

$$\left(\sum \delta W\right)_{AB} = \sum \vec{F}_{Ai} \cdot d\vec{r}_{Ai/O} + \sum \vec{F}_{Bj} \cdot d\vec{r}_{Bj/O} = 0 \qquad (9.25)$$

Equation (9.25) is the principle of virtual work for a system of rigid bodies with frictionless connections. Equation (9.25) implies that we only have to consider the virtual works done by the external forces, and their sum will be zero. We do not include the virtual works done by external moments in Equation (9.25). The work done by moments can be added to Equation (9.25) similar to Equation (9.22).

In fact, Equation (9.25) also provides other possibilities. For example, if we fix rigid body B, i.e. not allowing virtual displacement to rigid body B, then $d\vec{r}_{Bj/O} = 0$ and $d\vec{r}_c = 0$. In such a condition, Equation (9.24) is trivial, but Equation (9.23) becomes

$$\left(\sum \delta W\right)_A = \sum \vec{F}_{Ai} \cdot d\vec{r}_{Ai/O} = 0 \qquad (9.26)$$

Similarly, we can have

$$\left(\sum \delta W\right)_B = \sum \vec{F}_{Bj} \cdot d\vec{r}_{Bj/O} = 0 \qquad (9.27)$$

As a result, we can have three equations by applying the principle of virtual work differently.

We will use two examples from the textbook of Statics by Meriam and Kraige[2] for illustration.

For Fig. 9.17, a known force, P, is applied normal to the link as shown. We need to determine the force R required to maintain an equilibrium of the linkage at the angle θ. Note that this linkage system has only one degree of freedom (see Section 7.1.1, Chapter 7). If point C has a virtual displacement $d\vec{r}_C$, then point B will have a virtual displacement $d\vec{r}_B$ as shown in Fig. 9.17, and both displacements will be related. Graphically, it

---

[2]Meriam, J.L. and Kraige, L.G. (2011). *Engineering Mechanics: Statics*, 7th edition, ISBN-13: 978-0470917879.

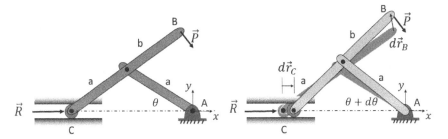

Figure 9.17:   Equilibrium condition for a linkage system (see footnote 2).

is difficult to determine how these two displacements are related. It would be easier if we simply use calculus to solve the problem. First, applying the principle of virtual work, we have

$$\sum \delta W = \vec{P} \cdot d\vec{r}_B + \vec{R} \cdot d\vec{r}_C = 0 \tag{9.28}$$

We then determine the force vectors and position vectors as

$$\vec{r}_B = (b - a)\cos\theta\vec{i} + (a + b)\sin\theta\vec{j} \tag{9.29}$$

$$\vec{r}_C = -2a\,\cos\theta\vec{i} \tag{9.30}$$

$$\vec{P} = P(\sin\theta\vec{i} - \cos\theta\vec{j}) \tag{9.31}$$

$$\vec{R} = R\vec{i} \tag{9.32}$$

Now, we can take derivative for Equations (9.29) and (9.30),

$$d\vec{r}_B = (a - b)\sin\theta\,d\theta\vec{i} + (a + b)\cos\theta\,d\theta\vec{j} \tag{9.33}$$

$$d\vec{r}_C = 2a\,\sin\theta\,d\theta\,\vec{i} \tag{9.34}$$

Introducing Equations (9.31–9.34) into Equation (9.28), we readily obtain $R = \frac{P}{2\sin\theta}(\frac{b}{a} + \cos 2\theta)$.

For the example in Fig. 9.18, we want to determine the equilibrium configuration when a horizontal force, $\vec{P}$, is applied at point E.

We will determine all the force and position vectors as in the previous example.

$$\vec{P} = -P\vec{i} \tag{9.35}$$

$$m\vec{g} = mg\vec{j} \tag{9.36}$$

$$\vec{r}_E = -(l\,\sin\theta_1 + l\,\sin\theta_2 + l\,\sin\theta_3)\,\vec{i} \tag{9.37}$$

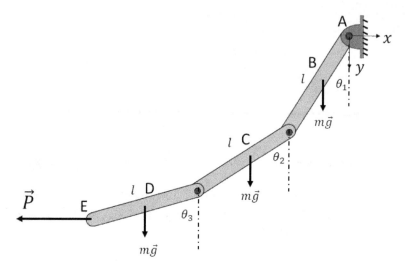

Figure 9.18: Equilibrium condition for a 3-link linkage system (see footnote 2).

$$\vec{r}_B = -\frac{1}{2} l \cos\theta_1 \, \vec{j} \tag{9.38}$$

$$\vec{r}_C = -\left(l \cos\theta_1 + \frac{1}{2} l \cos\theta_2 \right) \vec{j} \tag{9.39}$$

$$\vec{r}_D = -\left(l \cos\theta_1 + l \cos\theta_2 + \frac{1}{2} l \cos\theta_3 \right) \vec{j} \tag{9.40}$$

Correspondingly, we can define the potential energy of each link as

$$U_B = -\frac{1}{2} m g l \cos\theta_1 \tag{9.41}$$

$$U_C = -mg \left(l \cos\theta_1 + \frac{1}{2} l \cos\theta_2 \right) \tag{9.42}$$

$$U_D = -mg \left(l \cos\theta_1 + l \cos\theta_2 + \frac{1}{2} l \cos\theta_3 \right) \tag{9.43}$$

Let us first apply the virtual work principle to the entire system,

$$\sum \delta W = \delta W_P + \delta W_B + \delta W_C + \delta W_D$$

$$= \vec{P} \cdot \delta\vec{r}_E - \delta U_B - \delta U_C - \delta U_D = 0 \tag{9.44}$$

where

$$d\vec{r}_E = -(l \cos\theta_1 \, \delta\theta_1 + l \cos\theta_2 \, \delta\theta_2 + l \cos\theta_3 \, \delta\theta_3) \, \vec{i} \qquad (9.45)$$

$$dU_B = \frac{1}{2} mgl \sin\theta_1 \, \delta\theta_1 \qquad (9.46)$$

$$dU_C = mg \left(l \sin\theta_1 \, \delta\theta_1 + \frac{1}{2} l \sin\theta_2 \, \delta\theta_2 \right) \qquad (9.47)$$

$$U_D = mg \left(l \sin\theta_1 \, \delta\theta_1 + l \sin\theta_2 \, \delta\theta_2 + \frac{1}{2} l \sin\theta_3 \, \delta\theta_3 \right) \qquad (9.48)$$

Of course, Equation (9.44) is not very useful because there are too many unknowns. We can now apply the virtual work principle to just a part of the entire system by allowing only one virtual displacement at a time. First, keep $\delta\theta_1 = \delta\theta_2 = 0$ while $\delta\theta_3 \neq 0$, Equation (9.44) is then simplified to

$$\sum (\delta W)_{\theta_3} = Pl \cos\theta_3 \, \delta\theta_3 - \frac{1}{2} mgl \sin\theta_3 \, \delta\theta_3 = 0, \qquad (9.49)$$

which leads to $\tan\theta_3 = \frac{2P}{mg}$.

Similarly, we have

$$\sum (\delta W)_{\theta_2}$$
$$= Pl \cos\theta_2 \, \delta\theta_2 - mgl \sin\theta_2 \, \delta\theta_2 - \frac{1}{2} mgl \sin\theta_2 \, \delta\theta_2 = 0 \qquad (9.50)$$

with $\delta\theta_1 = \delta\theta_3 = 0$ while $\delta\theta_2 \neq 0$. We have $\tan\theta_2 = \frac{2P}{3mg}$. Finally, we have

$$\sum (\delta W)_{\theta_1} = Pl \cos\theta_1 \, \delta\theta_1 - mgl \sin\theta_1 \, \delta\theta_1$$
$$- mgl \sin\theta_1 \, \delta\theta_1 - \frac{1}{2} mgl \sin\theta_1 \, \delta\theta_1 = 0 \qquad (9.51)$$

with $\delta\theta_2 = \delta\theta_3 = 0$ while $\delta\theta_1 \neq 0$, we obtain $\tan\theta_2 = \frac{2P}{5mg}$.

Again, for solving this problem, it would be much easier to use math, instead of geometry, as we did with 3D force calculations in Chapter 2.

9.5 More Math for Stability Analysis

We now present a mathematical stability criterion based on the energy concept. We already learned that when an equilibrium is achieved, for a system subjected to only conservative forces, its total potential energy will be either at maximum or minimum, i.e. $dU_{\text{total}} = 0$. Note that the previous statement is not true if the system is subject to non-conservative forces, which do work. For such a conservative system, the total potential energy is a function of position, $U_{\text{total}} = U_{\text{total}}(x, y, z)$. For a system with multiple rigid bodies, the total energy will be a function of many positions, not just limited to three variables.

Let us learn a few new mathematical tools for this discussion. In basic calculus, the derivative is taken with respect to a scalar, such as dt in $\dot{x} = \frac{dx}{dt}$. We can take a derivative of a vector with respect to a scalar too, such as, $\frac{d\vec{r}}{dt} = \frac{dx}{dt}\vec{i} + \frac{dy}{dt}\vec{j} + \frac{dz}{dt}\vec{k}$. We do not get into the situation when $d\vec{i}/dt \neq 0$ where a rotating non-inertial reference frame is used. It would be a topic in Dynamics. Can we take a derivative with respect to a vector? To answer this question, we introduce some new notations so that we can deal with any number of dimensions, not just x, y, and z.

For a vector in an n-dimensional space, the unit vectors for each dimension are defined as $\vec{e}_i$, $i = 1, 2, \ldots, n$. The position vector of a point in this n-dimensional space is

$$\vec{q} = q_1\vec{e}_1 + q_2\vec{e}_2 + \cdots + q_n\vec{e}_n \tag{9.52}$$

Mathematically, we can introduce a row vector and a column vector so that

$$\vec{q} = q_1\vec{e}_1 + q_2\vec{e}_2 + \cdots + q_n\vec{e}_n = \vec{e}\boldsymbol{q} \tag{9.53}$$

where $\vec{e} = [\vec{e}_1\vec{e}_2 \ldots \vec{e}_n]$ and $\boldsymbol{q} = \begin{bmatrix} q_1 \\ q_2 \\ \vdots \\ q_n \end{bmatrix}$. It becomes troublesome to write down $\vec{e}$ every time. We will just use $\boldsymbol{q}$, which represents $\vec{q}$, while keeping it as a column vector. We will further simplify it and use q for this column vector instead of $\boldsymbol{q}$.

Now, let us take a derivative of a scalar function, U, with respect to q. This is called directional derivative and is defined as

$$\frac{\partial U}{\partial q} = \left[\frac{\partial U}{\partial q_1} \quad \frac{\partial U}{\partial q_2} \quad \cdots \quad \frac{\partial U}{\partial q_n} \right] \tag{9.54}$$

where an operator $\frac{\partial}{\partial q} = [\frac{\partial}{\partial q_1} \frac{\partial}{\partial q_2} \cdots \frac{\partial}{\partial q_n}]$ is defined, as a row vector. Therefore, if we take a directional derivative of a scalar, we end up with a row vector using the directional derivative operator. A directional derivative is a partial derivative. To get the total derivative, we need to use the total derivative operator,

$$\frac{\partial}{\partial q} dq = \begin{bmatrix} \dfrac{\partial}{\partial q_1} & \dfrac{\partial}{\partial q_2} & \cdots & \dfrac{\partial}{\partial q_n} \end{bmatrix} \begin{bmatrix} dq_1 \\ dq_2 \\ \vdots \\ dq_n \end{bmatrix}$$

$$= dq_1 \frac{\partial}{\partial q_1} + dq_2 \frac{\partial}{\partial q_2} + \cdots + dq_n \frac{\partial}{\partial q_n} \tag{9.55}$$

Applying the total derivative vector to U, we have

$$\frac{\partial U}{\partial q} dq = \begin{bmatrix} \dfrac{\partial U}{\partial q_1} & \dfrac{\partial U}{\partial q_2} & \cdots & \dfrac{\partial U}{\partial q_n} \end{bmatrix} \begin{bmatrix} dq_1 \\ dq_2 \\ \vdots \\ dq_n \end{bmatrix}$$

$$= \frac{\partial U}{\partial q_1} dq_1 + \frac{\partial U}{\partial q_2} dq_2 + \cdots + \frac{\partial U}{\partial q_n} dq_n \tag{9.56}$$

The total derivative of a scalar is still a scalar, as indicated by Equation (9.56).

To take second directional derivatives, such as $\frac{\partial^2 U}{\partial q_1 \partial q_2}$, we will need a new operator,

$$\left(\frac{\partial}{\partial q}\right)^T \left(\frac{\partial}{\partial q}\right) = \begin{bmatrix} \dfrac{\partial}{\partial q_1} \\ \dfrac{\partial}{\partial q_2} \\ \vdots \\ \dfrac{\partial}{\partial q_n} \end{bmatrix} \begin{bmatrix} \dfrac{\partial}{\partial q_1} & \dfrac{\partial}{\partial q_2} & \cdots & \dfrac{\partial}{\partial q_n} \end{bmatrix}$$

$$= \begin{bmatrix} \dfrac{\partial^2}{\partial q_1 \partial q_1} & \cdots & \dfrac{\partial^2}{\partial q_1 \partial q_n} \\ \vdots & \ddots & \vdots \\ \dfrac{\partial^2}{\partial q_n \partial q_1} & \cdots & \dfrac{\partial^2}{\partial q_n \partial q_n} \end{bmatrix} = \left(\frac{\partial^2}{\partial q \partial q}\right) \tag{9.57}$$

where T is the transpose operator to turn a column vector to a row vector or vice versa.

The second directional derivative of a scalar becomes

$$\left(\frac{\partial}{\partial q}\right)^T \left(\frac{\partial U}{\partial q}\right) = \left(\frac{\partial^2 U}{\partial q \partial q}\right) = \begin{bmatrix} \frac{\partial}{\partial q_1} \\ \frac{\partial}{\partial q_2} \\ \vdots \\ \frac{\partial}{\partial q_n} \end{bmatrix} \begin{bmatrix} \frac{\partial U}{\partial q_1} & \frac{\partial U}{\partial q_2} & \cdots & \frac{\partial U}{\partial q_n} \end{bmatrix}$$

$$= \begin{bmatrix} \frac{\partial^2 U}{\partial q_1 \partial q_1} & \cdots & \frac{\partial^2 U}{\partial q_1 \partial q_n} \\ \vdots & \ddots & \vdots \\ \frac{\partial^2 U}{\partial q_n \partial q_1} & \cdots & \frac{\partial^2 U}{\partial q_n \partial q_n} \end{bmatrix} \tag{9.58}$$

Note that the second directional derivative of a scalar is an n by n matrix.

Now, let us apply Taylor's expansion for a total potential energy function of a system with multiple rigid bodies with respect to an equilibrium position, q_0,

$$U = U_0 + \left(\frac{\partial U}{\partial q}\right)_{q_0} (q - q_0) + \frac{1}{2}(q - q_0)^T \left(\frac{\partial^2 U}{\partial q \partial q}\right)_{q_0} (q - q_0) + \text{(H.O.T.)} \tag{9.59}$$

We neglect the high-order terms (H.O.T.) in Equation (9.59) because they diminish rapidly as long as $|q - q_0|$ is very small.

Equation (9.59) clearly indicates that at equilibrium, $U = U_0$ as $q = q_0$ and $\left(\frac{\partial U}{\partial q}\right)_{q_0} dq = 0$ because $dU = 0$. Therefore,

$$U = U_0 + \frac{1}{2}(q - q_0)^T \left(\frac{\partial^2 U}{\partial q \partial q}\right)_{q_0} (q - q_0) + \cdots \tag{9.60}$$

When the system deviates slightly from q_0, the total potential energy will either increase or decrease depending on the property of the matrix $\left(\frac{\partial^2 U}{\partial q \partial q}\right)_{q_0}$.

For practical purposes, the matrix $\left(\frac{\partial^2 U}{\partial q \partial q}\right)_{q_0}$ is symmetric. The matrix $\left(\frac{\partial^2 U}{\partial q \partial q}\right)_{q_0}$ is positive definite if $\frac{1}{2}(q - q_0)^T \left(\frac{\partial^2 U}{\partial q \partial q}\right)_{q_0} (q - q_0) > 0$ for all

$q \neq q_0$. In this case, U_0 is a minimum. On the other hand, $\left(\frac{\partial^2 U}{\partial q \partial q}\right)_{q_0}$ is negative definite if $\frac{1}{2}(q - q_0)^T \left(\frac{\partial^2 U}{\partial q \partial q}\right)_{q_0} (q - q_0) < 0$ for all $q \neq q_0$. U_0 is a maximum. If $\left(\frac{\partial^2 U}{\partial q \partial q}\right)_{q_0}$ is neither positive definite or negative definite, it is indefinite. If $\left(\frac{\partial^2 U}{\partial q \partial q}\right)_{q_0}$ is zero (a zero matrix when every term is zero), then we will have to examine higher-order terms to determine the property of U_0.

Now, the question is: Does a stable equilibrium have a potential energy maximum or minimum? From Fig. 9.1, this question appears to be trivial and the answer should be "minimum", at least locally. However, Fig. 9.1 is only a case with gravity, and it is dangerous to generalize a simple case for conclusion, which could lead to serious mistakes.

To answer this seemingly simple question, let us consider it from a fundamental point of view. When an object deviates slightly from its equilibrium position in a space where only conservative forces are doing work, if this object can return to the stability position "naturally" by the action of the conservative forces, then the equilibrium is stable (case A in Fig. 9.1); otherwise, it is unstable (cases B and C in Fig. 9.1) or neutrally stable (case D in Fig. 9.1). We are only interested in the stable case here. Let us use our new mathematical tool for explanation.

The equilibrium position is q_0. The object is at position q, not too far from q_0. The displacement vector is $(q - q_0)$. In order to return to q_0, a resulting conservative force, Q, must be generated and the direction of this force must be in some way opposite to $(q - q_0)$ so that the object could be "pushed" back. Therefore, as shown in Fig. 9.19, we have

$$Q \cdot (q - q_0) = |Q||(q - q_0)| \cos \theta < 0 \tag{9.61}$$

In other words, the angle between Q and $(q - q_0)$ has to be larger than 90 degrees, that is, they are on the same side.

Recalling Equations (9.6), (9.8), and (9.10), we have

$$\delta W = -dU(r_1, r_2) = \vec{Q}(\vec{r}) \cdot d\vec{r} \tag{9.62}$$

Or, in the notation of directional derivatives,

$$Q = -\frac{\partial U}{\partial q} \tag{9.63}$$

Introducing Equation (9.63) into (9.61), we have

$$Q \cdot (q - q_0) = -\frac{\partial U}{\partial q}(q - q_0) < 0 \tag{9.64}$$

or

$$\frac{\partial U}{\partial q}(q - q_0) > 0 \tag{9.65}$$

In other words, along the direction $(q - q_0)$, $|\frac{\partial U}{\partial q}| > 0$. This means that the potential energy will increase when the object deviates from the equilibrium position.

Therefore, the potential energy at the equilibrium should be at least a local minimum, i.e. within a small region in the neighborhood of q_0, the potential energy is the lowest if the equilibrium is stable. This also indicates that for Equation (9.60), if we have

$$\frac{1}{2}(q - q_0)^T \left(\frac{\partial^2 U}{\partial q \partial q}\right)_{q_0} (q - q_0) > 0 \tag{9.66}$$

then the potential energy is a minimum and the equilibrium is stable. If $\frac{1}{2}(q - q_0)^T \left(\frac{\partial^2 U}{\partial q \partial q}\right)_{q_0} (q - q_0) = 0$, then we need to examine the higher-order term for the stability analysis.

If $\frac{1}{2}(q - q_0)^T \left(\frac{\partial^2 U}{\partial q \partial q}\right)_{q_0} (q - q_0) < 0$, then it is a maximum and the equilibrium is unstable.

We can also link the concept of the conservative force and Equation (9.66). If we move further away by dq from position q, what will be the resulting force change, dQ? Will it result in a force strong enough to push the object back to the equilibrium? If so, the stability is "stronger".

$$dQ = \frac{\partial Q}{\partial q} dq = -\left(\frac{\partial^2 U}{\partial q \partial q}\right) dq \tag{9.67}$$

and if

$$dq \cdot dQ = dq \cdot \frac{\partial Q}{\partial q} dq = -(dq)^T \left(\frac{\partial^2 U}{\partial q \partial q}\right) dq < 0 \tag{9.68}$$

we have a similar result of $(dq)^T \left(\frac{\partial^2 U}{\partial q \partial q}\right) dq > 0$. From Equation (9.68), we know dQ is on the same side of Q, as shown in Fig. 9.19. As a result, $|Q'| = |Q + dQ| > |Q|$. The interpretation is that if an object is away from the equilibrium position in a space of potential forces, the resulting potential force will push it toward the equilibrium position if the equilibrium

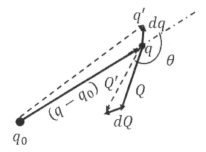

Figure 9.19: Equilibrium and stability analysis in a generalized space.

is stable. As long as the object is inside the stable region, being further away will result in a **stronger** force to push the object back to the equilibrium position. According to Fig. 9.12, if the object is at the right from the equilibrium position (when the spring is neither compressed or stretched), the spring is stretched. The resulting spring force will pull the object to the left back to the equilibrium position. If the object is further to the right, the resulting force is higher to pull it back. Finally, a little caution here: we do not consider the kinetic energy of the object here, which is a subject in Dynamics. If we do consider Dynamics, without friction, the object will end up oscillating around the equilibrium point; however, with friction, the object might not return to the equilibrium point when the spring is neither stretched nor compressed. Similarly, the stable equilibrium of case A in Fig. 9.1 is really not stable in a dynamic sense. Without friction, the object will roll back and forth around the equilibrium point.

Overall, stability analysis is a complex subject, and the discussion in this section only provides simple cases for a space where only potential energies are involved.

We will review a few stability analysis examples to illustrate how to use the analysis above. First, we will revisit the example of Fig. 9.14 with the total potential energy defined by Equation (9.11),

$$U_{\text{total}}(z) = U_k + U_g = \frac{1}{2}kz^2 + z\,mg\sin\theta \qquad (9.69)$$

where z is positive when the spring is stretched from the equilibrium position and is negative when the spring is compressed. The datum for the gravitational potential energy is defined at the equilibrium position. Because U_{total} is only a function of z, the derivative is a simple scalar derivative,

$$\frac{dU_{\text{total}}(z)}{dz} = kz + mg\sin\theta \tag{9.70}$$

Let $dU_{\text{total}}(z) = 0$, we find out the equilibrium position as $z_{eq} = -mg\sin\theta/k$, the same result as before and the spring is compressed. Now, we take the second derivative,

$$\frac{d^2 U_{\text{total}}(z)}{dz^2} = k > 0 \tag{9.71}$$

Therefore, this equilibrium position is stable.

Next, let us look at one example from the textbook of Meriam and Kraige (see footnote 2) for a more complicated case.

As shown in Fig. 9.20, each of the two uniform links has a mass M and a length a. The combined stiffness of the upper two springs is k, and same as the bottom two springs. When both bars are in the vertical position, the springs are neither stretched nor compressed.

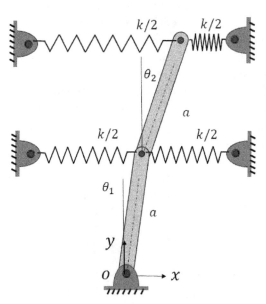

Figure 9.20: Equilibrium stability analysis example with multiple degrees of freedom (see footnote 2).

The total potential energy can be found out as

$$U_{\text{total}}(\theta_1, \theta_2) = U_k + U_g = \frac{1}{2}k(a\sin\theta_1)^2 + \frac{1}{2}k(a\sin\theta_1 + a\sin\theta_2)^2$$

$$+ \frac{1}{2}mga\cos\theta_1 + mga\left(\cos\theta_1 + \frac{1}{2}\cos\theta_2\right) \tag{9.72}$$

$$q^T = [\theta_1\ \theta_2] \tag{9.73}$$

$$\frac{\partial U}{\partial q} = \left[\frac{\partial U}{\partial q_1}\ \frac{\partial U}{\partial q_2}\right] = \left[\frac{\partial U}{\partial \theta_1}\ \frac{\partial U}{\partial \theta_2}\right] \tag{9.74}$$

$$\frac{\partial U}{\partial \theta_1} = ka^2\sin 2\theta_1 + ka^2\sin\theta_2\cos\theta_1 - \frac{3}{2}mga\sin\theta_1 \tag{9.75}$$

$$\frac{\partial U}{\partial \theta_2} = \frac{1}{2}ka^2\sin 2\theta_2 + ka^2\sin\theta_1\cos\theta_2 - \frac{1}{2}mga\sin\theta_2 \tag{9.76}$$

Letting $\frac{\partial U}{\partial \theta_1} = 0$ and $\frac{\partial U}{\partial \theta_2} = 0$, the equilibrium position is found to be $\theta_1 = 0$ and $\theta_2 = 0$.

Based on this equilibrium position, we found out that

$$\left(\frac{\partial^2 U}{\partial q \partial q}\right)_{0,0} = \begin{bmatrix} \dfrac{\partial^2 U}{\partial q_1 \partial q_1} & \dfrac{\partial^2 U}{\partial q_1 \partial q_2} \\ \dfrac{\partial^2 U}{\partial q_1 \partial q_2} & \dfrac{\partial^2 U}{\partial q_2 \partial q_2} \end{bmatrix}_{0,0} = \begin{bmatrix} 2ka^2 - \dfrac{3}{2}mga & ka^2 \\ ka^2 & ka^2 - \dfrac{1}{2}mga \end{bmatrix} \tag{9.77}$$

Let us first consider the case when $M = 10\,kg$, $k = 2000\,N/m$, and $a = 0.2\,m$. We have

$$A = \left(\frac{\partial^2 U}{\partial q \partial q}\right)_{0,0} = \begin{bmatrix} 130.57 & 80 \\ 80 & 70.19 \end{bmatrix} \tag{9.78}$$

If a symmetric matrix is positive definitive, then $trace(A) > 0$ and $det(A) > 0$. In this case, $trace(A) = 130.57 + 70.19 = 200.76 > 0$ and $det(A) = 130.57 \times 70.19 - 80 \times 80 = 2764.71 > 0$. This equilibrium position is stable.

If $M = 2\,kg$, $k = 20\,N/m$, and $a = 0.2\,m$,

$$B = \left(\frac{\partial^2 U}{\partial q \partial q}\right)_{0,0} = \begin{bmatrix} -4.286 & 0.800 \\ 0.800 & -1.162 \end{bmatrix} \tag{9.79}$$

Matrix B apparently is not positive definite because its trace is negative, i.e. $trace = (-4.286) + (-1.162) < 0$. To verify if B is negative definite, we can also determine its eigenvalues. Using matlab, we easily find that the eigenvalues of B are -0.9691 and -4.4789, both negative. Therefore, matrix B is negative definite, which means that the equilibrium is unstable.

In the above example, the equilibrium position is $\theta_1 = 0$ and $\theta_2 = 0$ if the springs are neither compressed nor stretched. If we stretch the lower left spring by b at the vertical position with the other springs the same condition as before, then the equilibrium position is no longer at $\theta_1 = 0$ and $\theta_2 = 0$.

The total potential energy becomes

$$U_{\text{total}}(\theta_1, \theta_2) = U_k + U_g = \frac{1}{4}k(a \sin \theta_1)^2 + \frac{1}{4}k(b + a \sin \theta_1)^2$$

$$+ \frac{1}{2}k(a \sin \theta_1 + a \sin \theta_2)^2$$

$$+ \frac{1}{2}mga \cos \theta_1 + mga \left(\cos \theta_1 + \frac{1}{2} \cos \theta_2 \right) \quad (9.80)$$

The corresponding equations for the equilibrium are

$$\frac{\partial U}{\partial \theta_1} = ka^2 \sin 2\theta_1 + \frac{1}{2}kab \cos \theta_1 + ka^2 \sin \theta_2 \cos \theta_1 - \frac{3}{2}mga \sin \theta_1 = 0 \quad (9.81)$$

$$\frac{\partial U}{\partial \theta_2} = \frac{1}{2}ka^2 \sin 2\theta_2 + ka^2 \sin \theta_1 \cos \theta_2 - \frac{1}{2}mga \sin \theta_2 = 0 \quad (9.82)$$

For the case with $M = 10\,kg$, $k = 2000\,N/m$, $a = 0.2\,m$, and $b = 0.02\,m$, the equilibrium position is found to be $\theta_1 = -1.164°$ and $\theta_2 = 1.327°$. A matlab script for solving the equilibrium positions is shown below. Readers can type in this simple script and execute it using matlab.

```
clear u v m a k g b
syms u v m k a g b
% u is theda1;
% v is theda2
% m is mass in kg
% a is length in m
% k is stiffness in N/m
% b is prestretching of the lower spring on the left side in m
g=9.81
m=10;
k=2000;
a=0.2;
b=0.02;
```

```
eq1 = k*a*sin(2*u) + k* a*sin(v)*cos(u)+ 0.5*a*b*k*cos(u) == (3*m*g/2)
* sin(u);
eq2 = 0.5*k*a*sin(2*v) + k*a*sin(u)*cos(v) == 0.5*m*g*sin(v);

eqns = [eq1,eq2 ];
S = solve(eqns);
sol = [S.u; S.v];
theda1 = sol(1);
theda1gree = theda1 *180.0 /3.141592654
theda2 = sol(2);
theda2degree = theda2 *180.0/3.141592654
```

We can go through the same verification and find out that the corresponding matrix

$$A = \left(\frac{\partial^2 U}{\partial q \partial q} \right)_{-1.164°,1.327°} = \begin{bmatrix} 130.48 & 80 \\ 80 & 70.19 \end{bmatrix},$$

which indicates that the new equilibrium position is also stable.

9.6 Concluding Remarks

In this chapter, we answered the question posed in the chapter title. The answer is NO! We introduced the concept of work, energy, conservative forces, and potential energy. We also discussed the method of virtual work to determine the equilibrium position without using the force equilibrium equations. Finally, we presented new mathematical tools to verify the stability of an equilibrium. It is important to point out that the stability analysis here is only valid for a system with conservative forces. If there are non-conservative forces, they should not do any work.

In Chapter 11, there are several problems related to the method of virtual work for practice, such as Problems 11.22, 11.27, 11.28, and 11.31.

In practice, if an equilibrium position is unstable, some control scheme can be devised to make it stable. For example, an inverted pendulum is unstable, but with a feedback control system it can be made stable. However, if the control system fails, the system becomes unstable again. Another example is related to airplanes. Without the control system, it will fall from the sky, as it is inherently unstable just like an inverted pendulum. The nuclear reactor of a nuclear power plant is also inherently unstable. We know of a few nuclear power plant failures that have caused major disasters. Therefore, when we think of the stability, we should RESPECT THE INSTABILITY! It is always preferred to design a system that is inherently stable. One final example is that of a conventional four-wheel vehicle; the car body is stable. However, vehicles on two wheels, such as motorcycles, bicycles, or Segways, are inherently unstable.

Chapter 10

A Case Study — Practical Sensor Design for Machining Force Monitoring

In this chapter, we present a case study of a sensor design via thorough force analysis for monitoring machining forces.[1] The machine involved is an instrumented machine tool spindle system with sensors from Promess Inc.[2] This spindle was donated by Ford Motor Company for supporting a bearing force thesis research.[3] Figure 10.1 illustrates the experimental setup of this research. In Fig. 10.2, the mounting of the spindle bearings is shown. The front bearings include a roller bearing and a pair of angular-contact ball bearings. To install the Promess sensor, a groove was ground on the outside surface of the outer race, and strain gauges were mounted inside the groove, as shown in Fig. 10.2.

In this chapter, we will also discuss how the force analysis we have discussed in this book can be applied to determine the entire sensing loop. The detailed bearings configuration is shown in Fig. 10.3. We will focus on the roller bearings. Each Promess sensor contains four strain gauges to form a Wheatstone bridge circuit. The vertical Promess sensor set shown in Fig. 10.4 is used for measuring the force in the vertical direction, while

---

[1]Tu, J.F. (1996). "Strain field analysis and sensor design for monitoring machine tool spindle bearing force". *International Journal of Machine Tools & Manufacture*, **36**(2): 203–216.

[2]https://www.promessinc.com/.

[3]Tu, J.F. (1991). "*On-line Preload Monitoring for High-Speed Anti-Friction Spindle Bearings Using Robust State Observers*". Ph.D. Dissertation: The University of Michigan, Ann Arbor, Michigan.

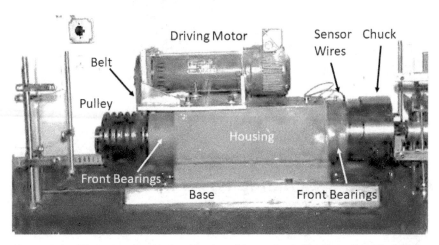

Figure 10.1: The setup of an instrumented machine tool spindle, donated by Ford Motor Company (see footnote 1).

Figure 10.2: The installation of the spindle bearings by a master technician, Mr. Stephen Wiley, of Main Street Motors, Ann Arbor, Michigan.

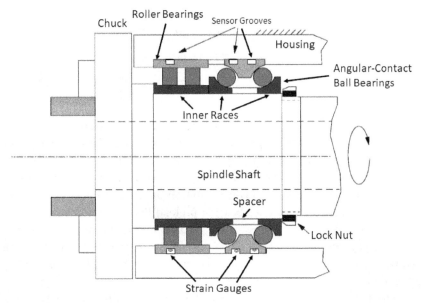

Figure 10.3: The front bearings configuration, including one set of roller bearings and a set of back-to-back angular-contact ball bearings (see footnote 1).

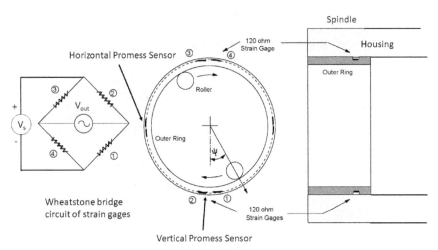

Figure 10.4: The Promess sensor mounting arrangement and the Wheatstone bridge (see footnote 1).

the horizontal Promess sensor set is used to measure the horizontal force. The combination of these two measurements can be used to measure forces in any directions.

10.1 Bearing Failures

Unexpected premature failures of machine tool spindle bearings are serious problems that could disrupt production, causing substantial financial losses due to loss of productivity and spindle rehabilitation cost. Many factors could contribute to premature bearing failures, such as excessive vibration, improper lubrication, external contamination, faulty installation, excessive external loading, and improper bearing preload. One common cause of the above factors is related to the internal contact force between the rolling elements and the bearing rings. It is conceivable that if this contact force can be measured, appropriate adjustments can then be made to ensure the designed fatigue life of the bearing.

The forces acting on a bearing include preload, preload changes due to thermal distortions in spindle (denoted as thermally induced preload), and the external load (see footnote 1). These forces change the contact conditions of the rolling elements. For example, when the bearing is subjected to preload only, the contact force acting on each rolling element is the same, resulting in an evenly distributed loading zone. On the other hand, if the bearing is subjected to a downward radial force without preload, only the lower portion of the rolling elements will carry the force, resulting in a partially loaded zone. The distributions of the contact force under more complicated loading conditions have been well documented.[4] A more detailed discussion on the force distribution is presented in Section 10.2.1.

In this chapter, a model is developed to analyze the strain field in the spindle bearing induced by the contact forces acting on the rolling elements. The strain model is developed based on classical solutions in elasticity with suitable assumptions. All the assumptions are justified by reasonable engineering judgments and experimental results. The contact force distribution can then be used to determine the corresponding preload and external forces.[5,6]

---

[4]Palmgren, A. (1959). *Ball and Roller Bearing Engineering.* SKF Industries, Inc.

[5]Tlusty, J. and Andrews, G.C. (1983). "A critical review of sensors for unmanned machining". *Annals of the CIRP.* **32**(2): 563–572.

[6]Promess Manual. (1984). *System Description, Tool Monitoring with Wear Control. Documentation No. 046: Turing I.*

10.2 Measurement Loop and Strain Model

In this section, a complete measurement loop model, which converts the force acting on the bearing to an electronic signal output, is developed. This measurement loop consists of the following five signal stages:

- External force;
- Internal bearing contact force;
- Stress distribution in outer ring/housing;
- Strain distribution;
- Electronic signal output.

For these five signal stages, four signal conversions are needed. A strain model is developed to fulfil the conversion from the force to the strain field. In what follows these four signal conversions are discussed in order.

10.2.1 Conversion #1: External force to internal force distribution

Without loss of generality, the signal conversions presented in this chapter are related to a precision lathe spindle supported radially by a double-row cylindrical roller bearing (SKF 3022 K/W 33). Figure 10.4 illustrates the configuration and the instrumentation of the bearing.

The bearing is assumed to be subjected to an external radial loading, $F_r = 1471.5\ N$ (the weight of the spindle). The bearing is assumed to be not preloaded; therefore, the loading is distributed to all the rollers in the lower half of the bearing. For bearings that are preloaded, the distribution will be different. The related bearing dimensions were: inside diameter, $D_i = 110\,\mathrm{mm}$; outside diameter, $D_o = 170\,\mathrm{mm}$; and the diameter of the spindle housing, $D_h = 220\,\mathrm{mm}$.

This loading distribution can be treated as a macro-version of the Hertzian contact problem, which states that the force acting on each individual roller is a function of respective deflection due to the loading. For an angular-contact ball bearing, the load distribution to each ball, subjected to both axial and radial loadings, is shown in Fig. 10.5, based on the classical book on bearings by Harris (1984).[7]

For a roller bearing, the contact angle $\alpha = 0$ and the axial load, $F_a = 0$. Assuming that the roller bearing is subjected to a radial force, F_r, with zero internal clearance and 50% loading zone, the force distribution can be

---

[7]Harris, T.A. (1984). *Rolling Bearing Analysis*, 2nd Edition. John Wiley and Sons: New York.

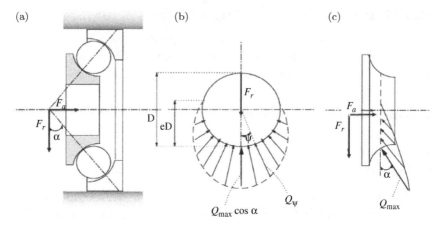

Figure 10.5:   Force distribution of an angular-contact ball bearing, subject to both radial and axial loadings (see footnote 1).

expressed as

$$Q_\Psi = Q_{\max} (\cos\Psi)^{1.11} \qquad (10.1)$$

where Q_Ψ is the force on the roller at angle Ψ, and $Q_{\max}$ is the force acting on the roller at $\Psi = 0$ and

$$Q_{\max} = \frac{F_r}{0.2453\, Z} \qquad (10.2)$$

where Z is the number of rollers and $Z = 30$ for the bearing of interest.

The vertical components of Q_Ψ will sum up to the radial force, F_r. We have

$$F_r = \sum_{\Psi=0}^{\Psi=\pm 90^\circ} Q_\Psi \cos\Psi \qquad (10.3)$$

Assuming a roller aligns with $\Psi = 0$, the relative angular positions of the rest of the rollers are at $\pm 12^\circ$, $\pm 24^\circ$, and so on. The distributed force at each roller becomes

$$Q_{\max} = 199.96\ N$$

$$Q_{\pm 12^\circ} = 195.11\ N$$

$$Q_{\pm 24^\circ} = 180.86\ N$$

$$Q_{\pm 36°} = 158.04 \ N$$

$$Q_{\pm 48°} = 128.01 \ N$$

$$Q_{\pm 60°} = 92.64 \ N$$

$$Q_{\pm 72°} = 54.30 \ N$$

$$Q_{\pm 84°} = 16.30 \ N \tag{10.4}$$

The above forces are needed to calculate the overall strain induced in the bearing and spindle system.

10.2.2 Contact condition

This section verifies whether or not a line contact assumption is valid for the above-distributed contact force, Q_Ψ. From an SKF bearing catalog, the static capacity of the roller bearing is found to be 87,100 N. This force is equally supported by two rows of rollers of length 20 mm. According to Equation (10.2), $Q_{max} = 5,918 \ N$; therefore, the Hertzian solution[8] predicts that, even under this maximum loading, the width of contact area is only 0.122 mm, which is substantially smaller than the outer ring diameter. Since this is the worst-case scenario, the resulting contact area nominal conditions will be narrower. Therefore, it is valid to assume line contact condition between the roller and the outer ring during normal operations. By verifying the line contact condition, the contact force at the roller can be considered as a concentrated force.

10.2.3 Conversion #2: Contact force to stress

We assume that the contact forces at each roller are at equilibrium because of the small mass of the roller. Therefore, the same contact forces we have determined by Equations (10.1–10.3) will be acting on the outer ring and housing.

The resultant stress field at the outer ring is assumed to be a linear combination of the stresses induced by all the contact forces, Q_Ψ. Therefore,

---

[8] Johnson, K.L. (1985). *Contact Mechanics.* Cambridge University Press, pp. 99–101; The relative curvature is $\hat{R} = 9.51$ mm.

we have the stress field function

$$\sigma(r,\theta) = \sum_{\Psi=0}^{\Psi=\pm180°} \sigma_\Psi(r, \theta, Q_\Psi) = \sum_{\Psi=0}^{\Psi=\pm180°} \sigma_0(r, \theta - \Psi, Q_\Psi) \qquad (10.5)$$

where $\Psi = \pm N \cdot 12°, N = 1, 2, \ldots$ and σ_0 is the stress due to a concentrated force at Ψ.

Because of symmetry, the stress field σ_Ψ at each angular position can be expressed in terms of σ_0 with an angle shift, as indicated in Equation (10.5). Therefore, we only need to derive σ_0.

There is no analytical solution directly applicable to predict the strain field of this bearing configuration shown in Fig. 10.4 because of the geometric complexity introduced by the groove ground around the outer ring for installing the Promess sensor. We need to find a way to determine the stress field under such unusual geometry.

In this section, we will introduce a new technique in the FBD analysis. In the earlier chapters, when we conduct the free-body diagram analysis, we always draw a boundary to separate a subsystem from the overall system. For this case study, we will show how to "fill in" an original system, instead of "cutting off" a subsystem. This "filling-in" technique was suggested by Professor Jim Barber of University of Michigan.

The original loading condition by a radial force, P_0, is shown in Fig. 10.6(a). We can draw a boundary to construct a free-body diagram as shown in Fig. 10.6(b). However, this case is very complicated because of the presence of the sensor groove. Today, with commercial finite-element method software, we can simulate the strain field of the entire system based on the exact geometry. However, we must conduct finite-element analysis carefully with experimental verifications because, without verifications, the results are prone to errors.

We will apply the "filling-in" technique so that we can simplify the FBD of Fig. 10.6(b). As shown in Fig. 10.6(c), we can add null pairs of forces around the groove. Adding these null pairs will not affect the overall loading condition. Then, we apply superposition to separate the null pairs, as shown in the enlarged view of the groove in Fig. 10.6(c). The second part of the separated forces can be made to be equivalent to the case when there is no groove in the ring. In other words, the groove is now "filled up". With this arrangement, we now have Fig. 10.6(d) with two loading conditions, conditions A and B, superimposed together. Condition A is a spindle system without the groove, subjected to the radial force, P_0, while condition B is a spindle system with the groove, subjected to the separate

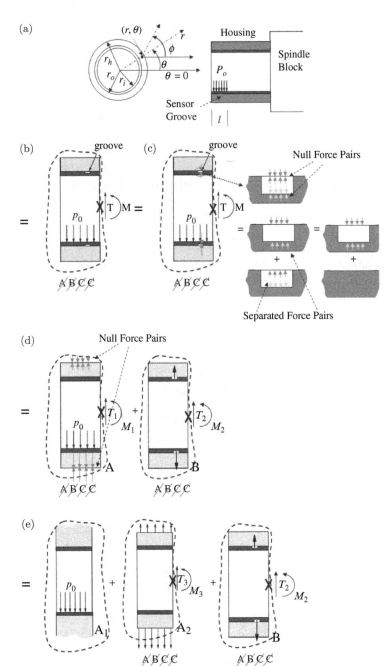

Figure 10.6:   Free-body diagram analysis with both "cutting-off" and "filling-in" techniques: (a) spindle housing with bearing outer ring; (b-e) FBDs with the filling-in and superposition method.

null forces. The stresses acting at the groove in loading condition B are determined from the stress distribution from condition A. For illustration purposes, we only show the null forces in the radial direction in Fig. 10.6(d).

Now, let us apply the same filling-in method, but this time to the outside of the housing. We will fill the housing to infinity. With that, we arrive at the FBD of Fig. 10.6(e), in which condition A is split into conditions A1 and A2. Condition A1 represents an infinitely thick plate with a hole, while condition A2 represents a hollow shaft subjected to loadings on the outside surface. Similarly, the loading on the outside surface in Case A2 is determined from Case A1 as the force at $R_h = 85\,\text{mm}$.

Now, we will discuss the stress for each condition of Fig. 10.6(e).

Case A1: This case is a 2D problem with a force P_0 acting at a hole in an infinite plate and no loading at infinity. Timoshenko and Goodier[9] derived an analytic solution for this problem, which gives the following stress function:

$$\Phi = -\frac{P_0}{\pi} \left\{ \phi\, r\, \sin\theta - \frac{1}{4}(1-\nu)r \log(r)\cos\theta \right.$$

$$\left. -\frac{1}{2}r\theta \sin\theta + \frac{d}{4}\log(r) - \frac{d^2}{32}(3-\nu)\frac{1}{r}\cos\theta \right\} \qquad (10.6)$$

where $P_0 = \frac{Q_0}{l}$ and l is the length of the roller.

Based on this stress function, three stress distributions can be derived as

$$\sigma_{rr} = \frac{1}{r}\frac{\partial \Phi}{\partial r} + \frac{1}{r^2}\frac{\partial^2 \Phi}{\partial \theta^2} \qquad (10.7)$$

$$\sigma_{\theta\theta} = \frac{\partial^2 \Phi}{\partial r^2} \qquad (10.8)$$

$$\tau_{r\theta} = \frac{\partial}{\partial r}\left(\frac{1}{r}\frac{\partial \Phi}{\partial \theta}\right) \qquad (10.9)$$

where σ_{rr} is the radial stress, $\sigma_{\theta\theta}$ is the circumferential stress, and $\tau_{r\theta}$ is the shear stress.

---

[9]Timoshenko, S.P. and Goodier, J.N. (1970). *Theory of Elasticity*, 3rd ed., McGraw-Hill, New York, ISBN-13: 978-0070701229.

A program written in the symbolic mathematical language of Mathe-matica (Appendix 10.A) is used to generate numerical solutions of the stress distributions (Appendix 10.B). The expressions of these distributions are not shown here because they are very lengthy. It should be noted that these stress distributions are all linear functions with respect to the contact force, Q_0. From the numerical results (Appendix 10.B), it is concluded that σ_{rr} is highly concentrated; the forces $Q_{\Psi \neq 0}$ have little effect on the radial stress field near $\theta = 0°$. The resultant circumferential stress $\sigma_{\theta\theta}$, on the other hand, is affected by all the contact forces. Finally, shear stress, $\tau_{r\theta}$, around $\theta = 0°$ is mainly affected by the two neighboring distributed forces.

Case A2: From the numerical solution of the external loading, the stress levels of all three stress distributions at $r = R_h = 110\,\mathrm{mm}$ are far less than the levels at $r = R_g = 83\,\mathrm{mm}$ (where the strain gauges are) based on the solution of Case A1. These stresses become the external loading of Case A2 at the housing, as shown in Fig. 10.6(e). The induced stress at $r = R_g = 83\,\mathrm{mm}$ due to the external loading at the housing will be much smaller than those induced in Case A1. Thus, the effect of Case A2 can be reasonably neglected.

Case B: The stresses acting at the surfaces of the groove are obtained from the stress distributions based on Case A1. At the inner surface, the stresses include σ_{rr} and $\tau_{r\theta}$ (see Appendix 10.B). The effects of these two stresses cannot be neglected for now, and they will be estimated in the next section.

10.2.4 Conversion #3: Stress to strain

The circumferential strain, $\epsilon_{\theta\theta}$, at $r = R_g$, picked by the gauge pair becomes

$$\epsilon_{\theta\theta}(R_g, \theta) = [\epsilon_{\theta\theta}]_{A1} + [\epsilon_{\theta\theta}]_B \tag{10.10}$$

which is a linear combination of the strains induced in Case A1 and Case B.

Strain in Case A1: The strain of Case A1, $[\epsilon_{\theta\theta}]_{A1}$, is calculated using the following equation:

$$[\epsilon_{\theta\theta}]_{A1} = \frac{1}{E}[(1 - \nu^2)\sigma_{\theta\theta} - \nu(1+\nu)\sigma_{rr}] \tag{10.11}$$

The resultant strain distribution is shown in Fig. 10.7, which indicates that, at the instant when a roller runs over the position $\Psi = 0$, a strain pulse is induced with a peak value about 6.5 μm/m at that position.

Note that although this peak value is underestimated, the nature of its concentrated distribution is creditable. The neighboring two rollers also produce their own pulses with slightly lower peak values centered at positions $\Psi = \pm 12°$. Since these three peak values are almost equal, they can be treated as a periodical wave function

$$[\epsilon_{\theta\theta}]_{A1} = f(\theta + n\theta_T), \qquad -20° < \theta < 20° \ and \ \theta_T = 12° \qquad (10.12)$$

Figure 10.7 also shows that the strain distribution is highly concentrated within $-2.5° < \theta < 2.5°$. Therefore, $[\epsilon_{\theta\theta}]_{A1}$ looks like a function of a series of moving waves as rollers run along the outer ring, while the peak position indicates the instantaneous roller position.

In Fig. 10.7, the location of the strain gauge pair is also shown, within $-5° < \theta < 5°$. The gauges are apart by a degrees, and each gauge occupies b degrees.

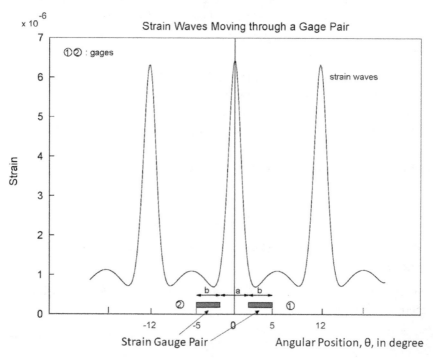

Figure 10.7: Induced strain waves in Case A1 w.r.t. the relative positions of the gauge pair (see footnote 1).

Strain in Case B: In Case B, the circumferential strain induced by both stresses, σ_{rr} and $\tau_{r\theta}$, at $r = R_g$ needs to be analyzed. Because of the groove at the outer ring, Case B is a quite complicated 3D problem and may need the Finite Element Method for accurate analysis. However, before venturing into the Finite Element Analysis, the upper bound of the circumferential strain induced by $\tau_{r\theta}$ at $r = R_g$ is first estimated to verify if there is a need for more complicated analysis. For the shear stress $\tau_{r\theta}$, the upper bound is estimated by simplifying Case B as a half plane with two uniform shear stresses acting along the inner surface of the groove. This assumption exaggerates the effect of the shear stress by several times. Under this assumption, the induced strain can be represented by two equal but opposite concentrated tangential forces, F_t, and the induced strain can then be estimated as[10]

$$\epsilon_x = \frac{4F_t(1 - \nu^2)}{\pi E x} = 2.5 \times 10^{-6} \text{ m/m} \tag{10.13}$$

Because the above upper bound strain level is less than one-half of the strain induced in Case A1 (Fig. 10.6(e)), it is safe to assert that the actual strain level is much smaller. As a result, it is decided to neglect the effect of $\tau_{r\theta}$, knowing the error will not exceed 33%. It will be shown later by experimental data that this assumption is reasonable.

The circumferential strain induced by σ_{rr} is also a complicated 3D problem. However, since the distribution of σ_{rr} is highly concentrated, it is reasonable to assume that the shape of the induced strain by this σ_{rr} will be highly similar to that of Case A1. Therefore, neglecting the strain induced by σ_{rr} does not incur significant qualitative error. This quantitative error can be corrected via sensor calibration.

Overall circumferential strain: Lastly, it is concluded that the circumferential strain, $\epsilon_{\theta\theta}$, picked up by the strain gauges is a periodic wave function, whose wave shape is mainly determined by the strain induced in Case A1 (Fig. 10.7). The peak value should be higher than the value shown in Fig. 10.7 because the effects of the three subcases have been neglected. It is also noted that the peak value of the function is linearly proportional

---

[10] Johnson, K.L. (1985). *Contact Mechanics*, Cambridge University Press, Cambridge, UK, p. 18, Equation (2.22a), ISBN 0-521-34796-3.

to the contact force, Q_0, and its frequency is related to spindle speed and the number of rollers.

10.2.5 Conversion #4: Strain to electronic signal

As shown in Figs. 10.4 and 10.7, the strain gauge pair is attached symmetrically to $\theta = 0°$ and forms a simple ladder circuit. As the rollers roll over the gauges, the gauges are subjected to fluctuating strain peaks, resulting in corresponding resistance changes. Since these strain waves can be treated as periodical waves, a moving coordinate attached on a roller can be chosen to "freeze" the strain pulses and let the gauge pair move through the "stationary" waves (Fig. 10.7).

With this moving coordinate, the resultant resistance changes of gauges #1 and #2 become two identical signals with a phase shift α.

$$\Delta R_1 = g(\omega t) = c_f \int_0^b \epsilon_{\theta\theta} R_g d\theta$$

$$\Delta R_2 = g(\omega t + \alpha) = c_f \int_\alpha^{b+\alpha} \epsilon_{\theta\theta} R_g d\theta \qquad (10.14)$$

where c_f is a conversion factor to convert from deformation to resistance, $\alpha = \frac{a+b}{12°} 360°$, and $\omega = \frac{360° \cdot 60}{RPM} \cdot Z$.

Note that the phase angle, α, and the angular positions, θ, are expressed in the same unit, but their physical meanings are completely different. Angular position θ gives a specific position on the outer race of the bearing, while the phase angle α represents the geometric relationship between ΔR_1 and ΔR_2.

The output signal from the gauge pair becomes

$$\Delta V = \frac{R + \Delta R_2}{2R + \Delta R_1 + \Delta R_2} V_s - \frac{1}{2} V_s$$

$$= \frac{\Delta R_2 - \Delta R_1}{2(2R + \Delta R_1 + \Delta R_2)} V_s$$

$$\cong \frac{\Delta R_2 - \Delta R_1}{4R} V_s$$

$$= \frac{g(\omega t + \alpha) - g(\omega t)}{4R} \qquad (10.15)$$

The output signal ΔV can be further amplified by an amplifier for signal transmission and post-signal processing.

10.2.6 Overall conversion: radial external force to electronic signal

Equations (10.1), (10.5–10.7), (10.11–10.12), and (10.15) describe the four signal conversions from external radial force to the voltage output of the strain gauge pair. They indicate that the amplitude of the voltage output is proportional to the radial force and its frequency is related to the spindle speed. In practice, a static calibration would be needed to establish a conversion curve between the force and the amplitude of the voltage output. This is also one of the main reasons that the qualitative accuracy of the sensing model is more important. With the above analysis, the sensor design can be optimized by selecting proper sensor sizes, locations, and configurations.

10.2.7 Optimal sensor design and experimental validation

With the overall conversion model of the Promess sensor, the sensor design can be further optimized by choosing a suitable phase shift, α. The objective is to make the signal ΔV (Equation (10.15)) smooth with least cross-over errors while maintaining sufficient signal amplitudes. Smaller cross-over errors are preferable because they simplify the post-signal processing and improve the precision. For instance, if the output wave is to be rectified by an analog circuit to a DC signal, an output with less cross-over errors will have less ripples in the rectified signal output. To minimize cross-over error, the length of the sensor has to be increased, but, at the same time, an increased gauge length reduces the amplitude (Equation (10.15)). It is obvious that a strain gauge with an infinitesimal size has a maximum sensitivity to convert the peak value of the strain distribution to an electronic signal. However, because of the finite size of strain gauges, only averaged strain will be converted, which reduces the amplitude of the output.

According to numerical simulation, the amplitude reduction of a gauge with an infinitesimal size to the one with a size of 6° is about 58%. If this reduction is not a concern and can be compensated electronically, then optimal design means selecting longer gauge size to reduce ripples.

If the sizes of the strain gauges are predetermined, the relative positions of the gauge pair need to be determined to obtain an appropriate phase, α, of Equation (10.15) so that the cross-over errors are minimized and the normal-mode signal level is maximized. The relative positions of the gauges are to be chosen in such a way that the converted signal is symmetrical and it preserves the peak amplitudes. From Equation (10.15), it is clear that if α is zero, i.e. ΔR_1 and ΔR_2 are in phase, the voltage output, ΔV, will be zero.

To have the most significant output, ΔR_1 and ΔR_2 should be separated by 180°. In other words, the selections of gage lengths "b" and gage gap "a" are subjected to a constraint:

$$\frac{a+b}{12°} \cdot 360° = 180° \tag{10.16}$$

Therefore,

$$a + b = 6° \tag{10.17}$$

As shown in Fig. 10.4, in the outer race of the bearing, the strain gages are to be placed within the region $-6° < \theta < 6°$ and symmetrical to $\theta = 0°$. For example, if the gage length is 4° (equal to 8.7 mm), then from Equation (10.14), the gap between the gages is 2° and gage #1 should be placed with its right edge aligned with $\theta = 5°$, while the left edge of gage #2 is aligned with $\theta = -5°$ (as shown in Fig. 10.4). A simulation of the voltage output based on this arrangement is shown in Fig. 10.8, which looks like a distorted sinusoidal signal with noticeable zero cross-over errors. This cross-over distortion is a feature of a composite signal combining two signals with a 180° phase shift, similar to the audio output of a B-B type (push–pull) stereo amplifier. The audio output of the B-B type amplifier is the composite of a positive half wave and a negative half wave. A gage size of 6° (8.70 mm) is the optimal design in terms of minimizing the cross-over distortion.

Figure (10.8) also shows a signal waveform measured by the strain gages mounted on the outer ring. Note that the simulated signal matches the measured waveform very well; both have noticeable distortions, unique for a compositions signal. The frequency of the strain waves is related to the spindle speed, and it will not affect the wave amplitude as long as the speed is not fast enough to induce contact resonance.

Other design concerns include the depth of the groove, the width of the groove, etc. From the solution of Equation (10.9), it is obvious that greater depth can induce higher strain. However, a deeper groove will weaken the strength of the outer ring. Therefore, the depth of groove should be determined considering both bearing safety and gage sensitivity. Another consideration of the groove depth is that a greater depth also induces more concentrated strain fluctuation and, thus, the effect of each roller may

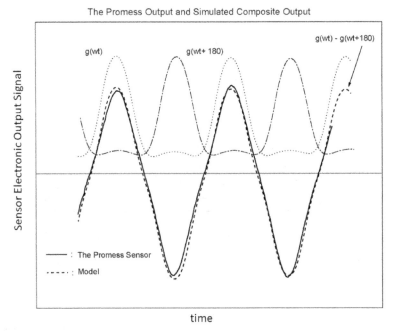

The Promess Output and Simulated Composite Output

Figure 10.8: Comparison between the simulated composite signal of the gage pair and the measured output of the Promess sensor (see footnote 1).

be treated as decoupled. The effect of the groove width is unclear from the proposed model since the problem is considered as a 2D problem. An engineering guess is that a greater width will provide a higher induced strain of Case B, but it may not be as sensitive as the groove depth. These issues may be investigated by using the Finite Element Analysis.

Another result from the model is that the absolute positions of the gages are not important. The gages can be placed anywhere within the region $-12° < \theta < 12°$ without significant loss of sensitivity as long as Equation (10.17) is maintained.

10.3 New Non-Invasive Sensing Scheme

From the above derivations, the lack of sensor sensitivity is one of the main reasons for going to great trouble to mount strain gages deep inside the spindle housing. This section explores the feasibility of using high-sensitivity strain gauges attached on the housing surface to monitor the spindle loading.

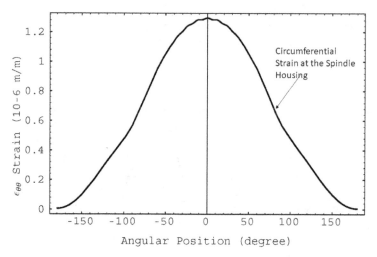

Figure 10.9: Circumferential strain field on the outside surface of the spindle housing (see footnote 1).

The strain field over the housing surface can be obtained by using similar but simpler analyses as discussed in the previous sections. The resulting circumferential, $\epsilon_{\theta\theta}$, strain field is shown in Fig. 10.9.

As illustrated in Fig. 10.9, the strain field on the housing surface is completely different from the one in Fig. 10.7. First, the distribution is no longer highly concentrated. Instead, the strain field is "mechanically averaged" by the thickness of the spindle housing. Second, the peak strain level is only about 1/6 the one at the bearing outer ring, calculated in load condition A1. The actual strain reduction can be larger when the effects of condition B are considered. With thicker housing (structurally stronger), more level reduction is expected. Therefore, the sensitivity of regular sensors is not sufficient. Furthermore, the sensor design discussed above is not suitable for this approach because the strain distribution is not highly concentrated. As a result, high-sensitivity sensors and a new sensor design are needed to ensure an adequate signal-to-noise ratio. Finally, since this strain field is rather "stationary", the waveform of the sensor output would be quite different from the one illustrated in Fig. 10.8.

With respect to this low-level and "dull" strain distribution, a different gage bridge is designed (Fig. 10.10). This new sensing scheme is denoted as a non-invasive sensing scheme here as opposed to the Promess scheme illustrated in Fig. 10.4. Compared with Figs. 10.4 and 10.7, the gage

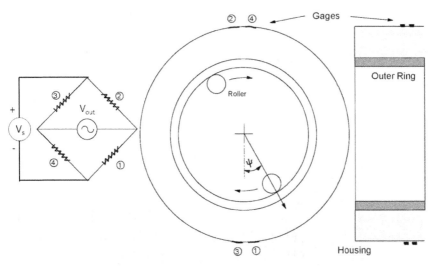

Figure 10.10: New non-invasive sensing scheme with strain gages mounted on the spindle housing outside surface (see footnote 1).

pair in Fig. 10.10 is placed on the opposite side of the spindle. For example, the paired gage #1 and gage #2 are placed at the bottom and at the top, respectively. This sensing scheme will behave as a DC-gain amplifier as opposed to the Promess scheme, which will modulate the output signal due to the bearing rotation. Even for a constant external force, the output of the Promess sensor is a modulated signal as shown in Fig. 10.7. If the external force is fluctuating at a higher frequency than the mechanical rotating motion of the rolling element, the Promess scheme would generate a signal output significantly different from the actual waveform of the force to be measured because of the modulation effect. Additional signal processing is required to decouple the modulation effect. On the other hand, no modulation is inherited by new sensing scheme with gages mounted on the housing; thus, it can reproduce the fluctuating force faithfully. As illustrated in Fig. 10.11, the waveform of the fluctuating force used for simulation is identical to the output of the housing mounted gages while the output of the conventional scheme becomes a modulation of these two different frequencies, as illustrated by the upper curve in Fig. 10.11.

The advantages of the new approach can be summarized as follows:

- Non-invasive;
- Simpler post-signal processing.

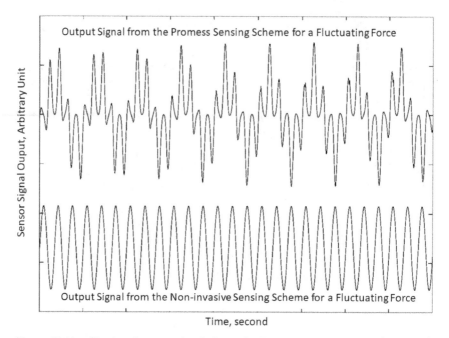

Figure 10.11:   Simulated output signals from the Promess sensing scheme (top curve) and from the new non-invasive scheme (bottom curve) for measuring a radial force with a sinusoidal fluctuation (see footnote 1).

The disadvantages are as follows:

- Needs high sensitivity sensors;
- May have lower signal-to-noise ratio.

If the radial force is large, the lower signal-to-noise ratio is likely not a problem.

This new non-invasive sensing scheme was perceived from the sensing model of the Promess sensing scheme. However, it is not really new because a patent[11,12] was filed by Boeing to measure the circumferential

---

[11] Jeppsson, J. (1989). *Method for Indicating End Mill Wear.* US Patent 4802095, Jan. 31, 1989.

[12] Jeppsson, J. (1999). "Sensor Based Adaptive Control and Prediction Software — Keys to Reliable HSM". At SME APEX'99 Machining and Metalworking Conference, Detroit, Sept 15.

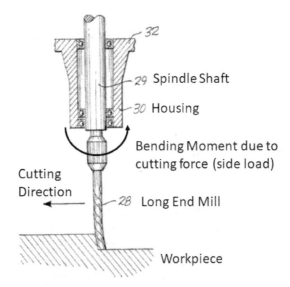

Figure 10.12: Bending of an end mill cutter due to the resultant side force (Jeppsson, 1989) (see footnote 11).

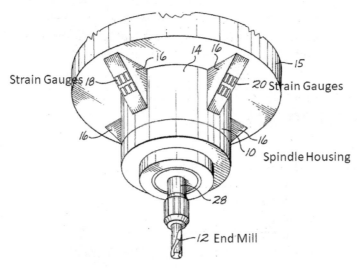

Figure 10.13: Strain-gauge-based bending moment sensor mounted on the spindle housing (see footnote 11).

strain at the spindle housing as a signal for the bending moment of the spindle. As shown in Figs. 10.12 and 10.13, this Boeing scheme Fig. 10.13) utilizes regular strain gages because the bending moment is very large when the spindle is used to machine aluminum stocks at high speeds. Readers can refer to the paper by Tu and Martin (2014) (see footnote 13) for more information on this application.

10.4 Concluding Remarks

In this chapter, we presented a case study on how the FBD analysis can be applied for practical sensor design. A new technique in the free-body diagram analysis, where we "fill" in materials, instead of "cutting" off materials, was introduced. This technique uses the principle of super-position for the "filling" in. Most importantly, this chapter shows how we can still reach a sound engineering analysis even though we do not have complete loading information of the system. With today's pervasive use of computer simulation, this chapter provides a solid, old-school way of force analysis, which is often needed to verify computer simulation results.

In this chapter, we also analyzed stresses and strains. Stresses are similar to the distributed forces we discussed in the previous chapters. The topics on stress and strain are related to the subject of mechanics of materials, which is a subsequent subject after Statics.

Appendix 10.A: *A Mathematica* Program Used to Generate Stress Fields

```
(* phi ......... stress function *)
(* P ........... point force      *)
(* theta ........ angle #1          *)
(* psi .......... angle #2         *)
(* r ........... radius            *)
(* d ........... diameter of the hole *)
(* pi ........... 3.14159          *)
(* mu .......... Poisson's ratio *)
```

---

[13] Jay, F.T. and Corless, M. (2014). "Review of sensor-based approach to reliable high speed machining at boeing — A tribute to Jan Jeppsson". *High Speed Machining*, 1: 1–17.

(* parameter values *)
(* point force in N/m; *)
(* P=200. /0.020 N/m; *)
(* Poisson's ratio: *) mu=0.3;
(* hole diameter: m   *) d=0.16;
pi=3.14159;
dummy=r Sin[theta] / (r Cos[theta] - d/2.);
psi=ArcTan[dummy];
phi1=- P / pi;
phi2= psi r Sin[theta];
phi3= - (1 -mu) r Log[r] Cos[theta] / 4.; phi4= - r theta Sin[theta]/2.;
phi5= d Log[r] /4.;
phi6= - d^2. (3. - mu) Cos[theta] /(32. r); phi=phi1 (phi2 + phi3 + phi4
+ phi5 + phi6);
(* d[h_]:= {D[h,x1],D[h,x2],D[h,x3], D[h,z]}; *)
phidtheta=   D[phi,theta];   phidr=D[phi,   r];   phidr2=D[D[phi,r],r];
phidr2=Simplify[phidr2];
(* sigmatt: circumferential stress *) sigmatt=phidr2;
phidtheta2= D[phidtheta,theta]; sigmarr=phidr /r + phidtheta2 / r^2 (*
sigmarr: radial stress *)
sigmarr=Simplify[sigmarr]; phidrtheta=D[phidtheta,r]
taortheta= phidtheta /r^2 - phidrtheta /r; (* sigmart: shear stress *)
sigmart=Simplify[taortheta];

Appendix 10.B: Calculated Stress Fields

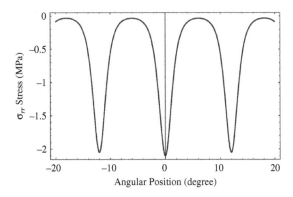

Figure B.1:   Radial stress, σ_{rr}, at the bearing outer ring.

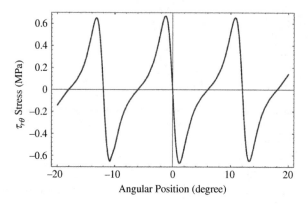

Figure B.2: Shear stress, $\tau_{r\theta}$, at the bearing outer ring.

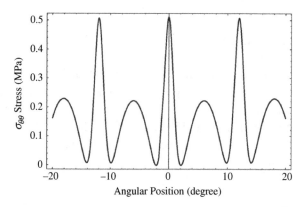

Figure B.3: Circumferential stress, $\sigma_{\theta\theta}$, at the bearing outer ring.

Chapter 11

Difficult Statics Problems Better Solved

In this chapter, we will solve classical problems in Statics using the methodologies discussed in Chapters 1–10. These problems are based on the review problems from *Statics* by Meriam, second edition.[1]

Problem 11.1: A person has two scales, one limited to 500 N (scale A) and one limited to 100 N (scale B). This person made a clever pulley system, as shown in figure (a), to measure his weight. He discovered that if he pulls the string hard and scale B shows 90 N, scale A will show 340 N. What is the actual mass of this person? Also, if he pulls harder and scale B shows 110 N, what will be the reading on scale A?

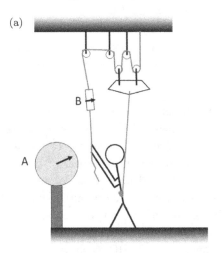

(a)

[1]Meriam, J.L. (1975). *Statics*, 2nd Edition. John Wiley & Sons, Inc., pp. 345–356.

Solution: Before solving the problem, we need to relate the scale readings to the forces. Apparently, the reading of scale B indicates the tension of the string, which runs through the pulley system. The tension is the same along the entire string. On the other hand, the reading of scale A indicates the contact force between the feet of the person and the scale platform.

The weight of the person is related to the gravitational force acting on the person.

Step 1: Which FBD?

We need to first decide on how to construct an FBD so that it contains the forces related to those given and those to be found. The obvious choice of the FBD is shown in (b).

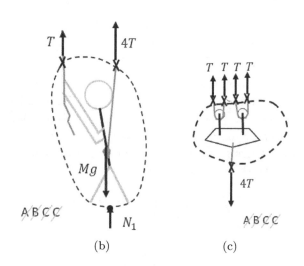

(b)   (c)

We go around the boundary and mark a contact point at the feet and two constraints at the strings. Then, we write down ABCC to account for all possible forces. There are no applied forces (cross out A). There is a body force. We mark Mg at the mass center of the person (cross out B). We have one contact point, and we draw a compressive force (pointing to the boundary) and mark it as N_1 (cross out the first C). Finally, we have two constraints. We know the tension on the string pulled by the hand is T. However, we do not know the tension at the strap tied to the body. For that, we can draw a second FBD as shown in (c). From (c), we easily find that the tension at the body strap should be $4T$. These forces are in tension; therefore, they are pointing away from the boundary.

Step 2: Governing equations
Based on the FBD of (b), we have

$$T + 4T - Mg + N_1 = 0 \qquad (11.1a)$$

Note that the force pointing upward is defined as positive. We know $T = 90\,\text{N}$ and $N_1 = 340\,\text{N}$

Step 3: Solving the governing equations

Counting unknowns: Before solving Equation (11.1a), we need to count the unknowns involved. There is only one unknown, M. Therefore, we can readily solve for it.

Solving for unknowns:

$$Mg = T + 4T + N_1 = 90 + 360 + 340 = 790 \qquad (11.1b)$$

$$M = \frac{790}{9.81} = 80.53\,(\text{kg}) \qquad (11.1c)$$

For the second question, now N_1 is an unknown, and so we have

$$110 + 440 - 80.53 \times 9.81 + N_1 = 0 \qquad (11.1d)$$

$$N_1 = 240\,(\text{N}) \qquad (11.1e)$$

Problem 11.2: A simulator as shown in (a) is constructed for studying human locomotion at reduced gravity. The feel of gravity is represented by the normal force felt by the person at the platform. Determine the angle θ so that the simulator can emulate the gravitational field of the moon. Note that the gravity on the moon is $1/6$ that of the earth. Will the friction of the platform affect the angle? What is the amount of force P?

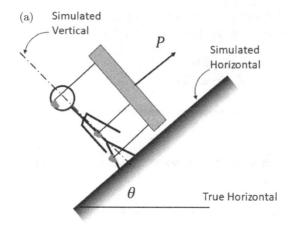

Solution:

Step 1: Which FBD?
Basically, we need to construct an FBD that will include the normal force at the feet. The obvious choice is shown in (b).

We can proceed to construct the full FBD. The force P is considered as an applied force. Going through the ABCC method, we complete the FBD of (b).

Step 2: Governing equations
The FBD of (b) is a 2D force condition. We should choose a reference coordinate so that the governing equation will be the simplest. With this consideration, we choose $\hat{x}$ and $\hat{y}$, aligned with the simulated horizontal and vertical.

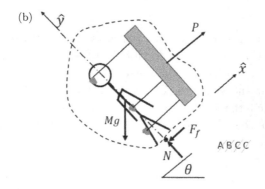

$$\left[\left(\sum F_i\right)_{\hat{x}} = 0\right] \quad P - mg\sin\theta - F_f = 0 \qquad (11.2a)$$

$$\left[\left(\sum F_i\right)_{\hat{y}} = 0\right] \quad -mg\cos\theta + N = 0 \qquad (11.2b)$$

As specified, we know $N = \frac{1}{6}mg$.

Step 3: Solving the governing equations

Counting unknowns: Before solving Equations (11.2a) and (11.2b), we need to count the unknowns involved. There is only one unknown, θ, involved in Equation (11.2b) and we can readily solve for it.

Solving for unknowns:
Solving Equation (11.2b), we have

$$-mg\cos\theta + \frac{1}{6}mg = 0 \qquad (11.2c)$$

$$\cos\theta = \frac{1}{6} \qquad (11.2d)$$

$$\theta = 80.41° \qquad (11.2e)$$

This angle is not affected by the friction between the feet and the platform. With the angle θ known, there are two unknowns in Equation (11.2a). We could not solve for P and F_f. We need to find another equation. Naturally, we might try to find an equation related to F_f. However, because this is at static equilibrium, we have an inequality, not an equation,

$$F_f \le \mu_s N = \mu_s \left(\frac{1}{6}mg\right) \qquad (11.2f)$$

In other words, the value of F_f can be any value from zero to $\frac{1}{6}\mu_s mg$. From the design point of view, we need to make sure that the simulator has enough capacity to provide the pulling force P. The maximal force needed will be when $F_f = 0$. Under this condition, we find out

$$P_{\max} = 0.986\,mg \qquad (11.2g)$$

Problem 11.3: Determine the location of the mass center of the circular wire with a radius r, as shown in (a).

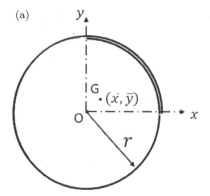

(a)

Solution: Based on Equation (4.2), we have

$$\int_0^{\frac{5}{4}\pi} f(x)(x-\bar{x})dx = 0 \tag{11.3a}$$

where $f(x)$ is the distributed force due to gravity by assuming that we place the circular wire vertically as shown in (b).

(b)

Based on (b), by introducing proper expressions in Equation (11.3a), we have

$$\int_0^{\frac{5}{4}\pi} f(x)xdx = \bar{x}\int_0^{\frac{5}{4}\pi} f(x)dx = \bar{x}\,Mg = \frac{5}{4}\pi r\rho g\bar{x} \tag{11.3b}$$

The left-hand side of the above equation is

$$\int_0^{\frac{5}{4}\pi} f(x)xdx = \int_0^{\frac{5}{4}\pi} r\cos\theta\,\rho grd\theta = r^2\rho g\sin\theta\Big|_0^{\frac{5}{4}\pi} = \frac{1}{2}r^2\rho g \tag{11.3c}$$

We then have $\bar{x} = \frac{2r}{5\pi}$. As the wire is symmetric, we have $\bar{y} = \frac{2r}{5\pi}$.

Problem 11.4: A band saw is shown in (a). The tension of the saw blade is maintained by a spring. During cutting, the cutting force to shear a tooth is 250 N, assuming the force is on one tooth. The static friction coefficient between the top pulley and the band saw is 0.4. Neglect the mass of the top pulley. The spring should be designed so that when the cutting force is too large, reaching 250 N, the band saw should slip over the top pulley, instead of shearing the blade tooth. Determine the appropriate spring tension for such a design.

(a)

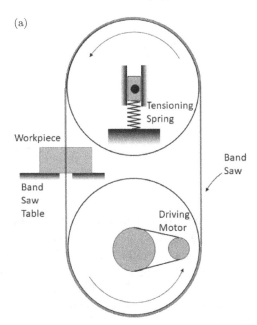

Solution: It is not clear how this problem can be solved. We will just go through the force analysis and find a solution along the way. First, we do an FBD as shown in (b) for the upper pulley and cut the band saw above where it engages the workpiece.

(b)

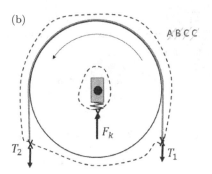

Based on (b), we have

$$F_k - T_1 - T_2 = 0 \tag{11.4a}$$

Equation (11.4a) is not too useful because we have three unknowns. We then construct a second FBD as shown in (c). Finally, we construct a third FBD as shown in (d) for a piece of band saw at the cutting point.

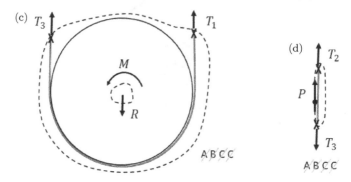

The lower pulley is the driving pulley. If the band saw is going to slip, it will slip at the upper pulley. The question is what will be the band saw tension when it slips? If it does not slip, we know from Equation (4.8) that the band saw tension will grow exponentially. It is safe to assume that if it slips, the change will be much less. For the first approach, we assume that once it slips at the upper pulley, the band saw tension at both sides of the upper pulley will be the same,

$$T_2 = T_1 \tag{11.4b}$$

Therefore, Equation (11.4a) becomes

$$F_k = 2T_1 \tag{11.4c}$$

If we can figure out T_1, then we can find out F_k.

Now, looking at (c), we can invoke Equation (4.8) of Chapter 4 regarding band saw tension,

$$T_3 = T_1 e^{\mu\theta} = T_1 e^{0.4\pi} = 3.5136\,T_1 \tag{11.4d}$$

From (d), we have

$$T_2 + P - T_3 = T_1 + 250 - 3.5136\,T_1 = 0 \tag{11.4e}$$

We obtain $T_1 = 99.46\,\text{N}$, and finally $F_k = 198.92\,\text{N}$.

This result indicates that this value of F_k is the minimal required for the upper pulley to slip without friction. If there is friction between the upper pulley and the band saw, then $T_2 > T_1$; thus, the value of F_k will increase.

On the other hand, if we keep the condition of Equation (11.4b) and $F_k = 198.91\,\text{N}$, and let, for example, $P = 125\,\text{N} < 250\,\text{N}$, then $T_2 = 3.5136\,T_1 - 125$ from Equation (11.4e). Invoking Equation (4.8) again for the upper pulley, we have

$$T_2 = T_1 e^{\hat{\mu}\pi} = 3.5136\,T_1 - 125 \tag{11.4f}$$

Finally, from Equation (11.4a), we have

$$198.92 - T_1 - T_1 e^{\hat{\mu}\pi} = 0 \tag{11.4g}$$

Solving Equations (11.4f) and (11.4g), we have $\hat{\mu} = 0.182$, which indicates no slip. In other words, after the spring force is set, the belt will not slip until the machining force reaches $P = 250\,\text{N}$.

For the same assumption and condition with $= 250\,\text{N}$, if we set the spring force lower, for example, at $F_k = 100\,\text{N}$, what would happen? From Equation (11.4c), we have $T_1 = \frac{F_k}{2} = \frac{100}{2} = 50\,(\text{N})$. From Equations (11.4d) and (11.4e), we have $\hat{\mu} = 0.57 > 0.4$, which exceeds the static friction limit. As a result, both upper and lower pulleys will slip.

Problem 11.5: Determine the forces at pin support A for the structure with two angled members pinned at point B.

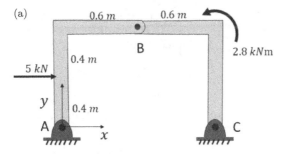

Solution: Let us first construct an FBD for the entire structure as shown in (b). When we go through the ABCC list, we have two applied forces (A) and two constraints (second C). Altogether, we have four unknowns in (b). We know that we cannot solve for all of them because we only have three equilibrium equations for a 2D problem.

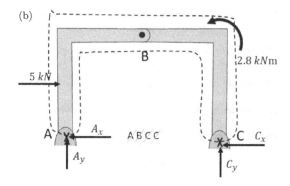

We will still write down the equilibrium equations and see if we can solve some of these unknowns.

$$\left[\sum F\right]_x = 0 \Rightarrow 5000 - A_x - C_x = 0 \qquad (11.5a)$$

$$\left[\sum F\right]_y = 0 \Rightarrow A_y + C_y = 0 \qquad (11.5b)$$

$$\left[\sum M\right]_C = 0 \Rightarrow -5000 \times 0.4 - 1.2A_y + 2800 = 0 \qquad (11.5c)$$

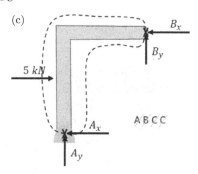

Luckily, Equation (11.5c) has only one unknown and we can solve it to obtain $A_y = 666.67\,\text{N}$. Then from Equation (11.5b), we have $C_y = -666.67\,\text{N}$. However, we cannot solve Equation (11.5a). We need to construct another FBD as (c). We have

$$\left[\sum F\right]_x = 0 \Rightarrow 5000 - A_x - B_x = 0 \qquad (11.5d)$$

$$\left[\sum F\right]_y = 0 \Rightarrow A_y + B_y = 0 \qquad (11.5e)$$

$$\left[\sum M\right]_B = 0 \Rightarrow 5000 \times 0.4 - 0.6A_y - 0.8A_x = 0 \qquad (11.5f)$$

Because we already know $A_y = 666.67\,N$, we can solve Equation (11.5f) for $A_x = 2000\,N$. The total force acting on pin A becomes $A = \sqrt{666.67^2 + 2000^2} = 2108\,N$.

Problem 11.6: The truss system in (a) is made of 45-degree right triangles. Determine the forces in members CD, DE, and CE, subjected to a loading P.

(a)

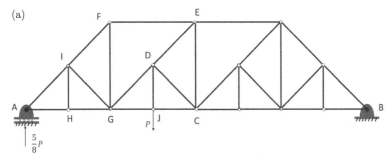

Solution: First, we readily find the vertical force at A as $\frac{5}{8}P$. For solving truss system problems, we have two approaches as discussed in Chapter 7.1.3. In the first approach, we go from the support and move to each joint, one at a time. We just have to make sure that at each joint, we can only have two unknown members. For this problem, we can go with this sequence A-H-I-F-G-J-D-E to find the desired solutions. To save time, we see if we can apply the second approach to draw a larger FBD, but we can only cut through three unknown members each time. If we do an FBD as shown in (b), we only cut three members, FE, GD, and GJ. From there, if we follow sequence J-D-E, we can solve the problems.

(b)

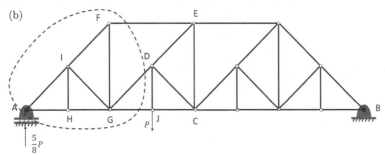

The answers are $CD = P/\sqrt{2}$ (compressive), $DE = \sqrt{2}P/8$ (compressive), and $CE = L/8$ (tensile). Readers are advised to actually carry out the calculations.

Problem 11.7: With respect to (a), at full load, $P = 4200\,\text{N}$. Determine the compression force at each leg of the equilateral frame ABC. The frame is used to distribute the vertical force equally.

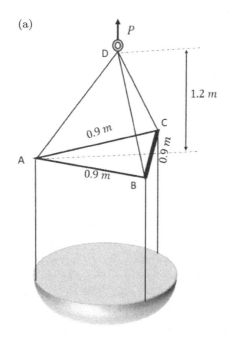

(a)

P

D

$0.9\,m$

C

1.2 m

A

$0.9\,m$

B

0.9 m

Solution: Based on the problem statement, we can easily deduce that the tension of each vertical line is $\frac{1}{3}P = 1400\,\text{N}$.

Now let us construct an FBD around point A, as shown in (b). We have four constraints. Two of them are the wire constraint and two of them are the cuts through the frame. Let us first make an assumption that the constraints at the frame will have only a compressive force along the frame leg. We will review this assumption later. With this assumption, when we go through the ABCC procedure, there are no applied forces (A), no body forces (B), and no contact forces (C), but four constraint forces (C). Because this is a 3D problem, it is advised to use math to solve the problem because, graphically, it would be too difficult to draw all the forces cleanly and correctly.

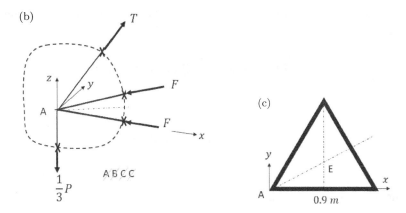

To use the vector analysis to solve the problem, we will define a suitable reference coordinate with respect to point A, shown in (b) and (c). From there, we determine the position coordinates of points A, B, C, D, and E.

We have

$$A : [0, 0, 0], \ B : [0.9, 0, 0], \ C : [0.9 \cos 60°, \ 0.9 \sin 60°, 0],$$

$$D : [0.45, 0.3 \sin 60°, 1.2], \ E : [0.45, 0.3 \sin 60°, 0] \tag{11.7a}$$

From these position coordinates, we can define the unit vectors in the force directions as

$$\vec{e}_{AB} = \vec{i}; \ \vec{e}_{AC} = \frac{\overrightarrow{AC}}{|\overrightarrow{AC}|} = 0.5\vec{i} + 0.866\vec{j};$$

$$\vec{e}_{AD} = \frac{\overrightarrow{AD}}{|\overrightarrow{AD}|} = 0.344\,\vec{i} + 0.199\vec{j} + 0.917\vec{k} \tag{11.7b}$$

From Equation (11.7b), we can now represent the force vectors as

$$\vec{T}_{AD} = T(0.344\,\vec{i} + 0.199\vec{j} + 0.917\vec{k}) \tag{11.7c}$$

$$\vec{F}_{AB} = -F\vec{i} \tag{11.7d}$$

$$\vec{F}_{AC} = -F(0.5\,\vec{i} + 0.866\vec{j}) \tag{11.7e}$$

$$\vec{F}_{wire} = -1400\vec{k} \tag{11.7f}$$

Using these force vectors, we simply need to construct the force equilibrium equations as

$$\left[\sum F\right]_x = 0 \Rightarrow 0.344T - F - 0.5F = 0 \tag{11.7g}$$

$$\left[\sum F\right]_y = 0 \Rightarrow 0.199\,T - 0.866\,F = 0 \tag{11.7h}$$

$$\left[\sum F\right]_z = 0 \Rightarrow 0.917\,T - 1400 = 0 \tag{11.7i}$$

Easily, we obtain $F = 350.3\,\text{N}$. Note that Equations (11.7g) and (11.7h) are identical, which is a check for errors.

(d)

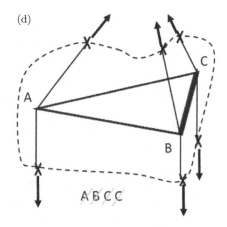

However, how do we know the assumption we made that the frame constraint only involves a compressive force along the frame leg? Note that the frame legs are welded together as a solid piece. As a result, when we draw a boundary across legs AB and AC, we are cutting through solid legs. Unlike the pined-pined rod constraint in a truss system, the constraint is a full constraint, involving forces in all three directions and moments. However, if we look closely at the forces acting on the frame, as shown in (d), we realize that the forces only act on points A, B, and C. For example, the force on points B and C cannot have any vertical components because they would cause moments with respect to point A. As a result, the resultant force on each point A, B, and C will be horizontal only. In other words, the frame ABC acts like a basic truss system, which justifies the assumption that the force on each leg can only be acting along the leg direction similar to the pined-pined rod constraint in the truss system.

Problem 11.8: A device to limit the cable tension consists of the toggle-operated jaw that shears the pin A when the cable tension exceeds a predetermined value. The pin is designed to shear at 12000 N. Determine the maximum tension T allowed by this device and the force acting on pin B.

(a)

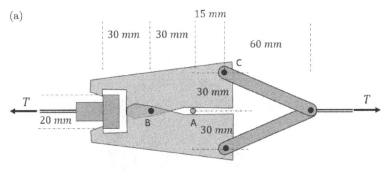

Solution: For this problem, we can construct an FBD as in (b). The forces at pin C can be easily found as the pined-pined rod constraint. Based on (b),

$$\left[\sum M\right]_B = 0 \implies F_A \cdot 30 - \frac{1}{4}T \cdot 45 - \frac{1}{2}T \cdot 30 + \frac{1}{2}T \cdot 10 = 0 \quad (11.8a)$$

(b)

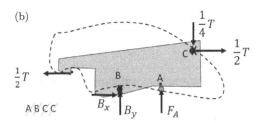

Solving Equation (11.8a), we have $T = 16941\,\text{N}$. Now from the force equilibrium,

$$\left[\sum F\right]_x = 0 \Rightarrow B_y - \frac{1}{4}T + 12000 = 0 \quad (11.8b)$$

$$\left[\sum F\right]_y = 0 \Rightarrow B_x - \frac{1}{2}T + \frac{1}{2}T = 0 \quad (11.8c)$$

Solving them, we have $B_y = -7765\,\text{N}$ and $B_x = 0\,\text{N}$.

The negative sign for B_y indicates that the actual force should be pointing downward, not upward as shown in (b).

Problem 11.9: A logger's hoist is activated by two hydraulic cylinders. For the particular position shown in (a), the oil pressure is found to be 1.2 MPa with the cylinder piston cross-section area $0.015\,m^2$. The log has a mass 2500 kg. Determine the force supported by pin C.

(a)

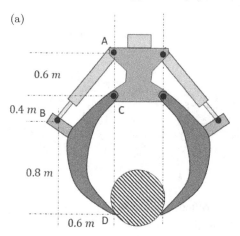

Solution: With an FBD as (b), we can have three force equilibrium equations,

$$\left[\sum F\right]_x = 0 \Rightarrow C_x - D_x - P\cos\theta = 0 \tag{11.9a}$$

$$\left[\sum F\right]_y = 0 \Rightarrow C_y - D_y - P\sin\theta = 0 \tag{11.9b}$$

$$\left[\sum M\right]_B = 0 \Rightarrow -1.2D_x - 0.4\,P\cos\theta + 0.6\,P\sin\theta = 0 \tag{11.9c}$$

where $\cos\theta = 0.6$, $\sin\theta = 0.8$, and

$$P = 1.2 \cdot 10^6 \cdot 0.015\,(N). \tag{11.9d}$$

There are four unknowns in the above equations. However, Equation (11.9c) involves just one unknown, D_x, which can be solved. We have $D_x = 3600\,N$. Now, Equations (11.9a) and (11.9b) have three unknowns. We can determine D_y by drawing an FBD for the log alone (not shown). From there, we easily find that $D_y = \frac{1}{2}M_{\log}g = 12262.5\,N$. Now, we can solve Equations (11.9a) and (11.9b). We have $C_x = 14400\,N$ and $C_y = 26663\,N$. The total force at C becomes $30302\,N$.

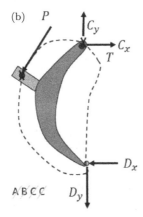

Note that the cylinder can be considered as a pined-pined rod; therefore, P is along the cylinder direction.

Problem 11.10: A uniform bar with negligible width is held against a pin (O) and a moment, M, is applied so that the bar could rotate around pin O, as shown in (a). The static friction coefficient between the bar and the surface is μ_s. Derive a relationship of M as a function of μ_s. The mass of the bar is m.

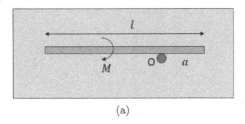

(a)

Solution: To solve this problem, we first assume that the normal distributed contact force between the bar and the surface is uniform. This makes the analysis of the frictional force possible. Otherwise, one has to conduct an experiment to find out the actual contact force distribution as in the case of the contact condition between a tire and the road.

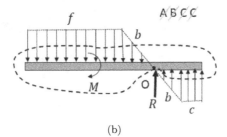

(b)

With this assumption, we construct an FBD as shown in (b). Remember that when the bar is not moving, the relationship between the static friction and the normal contact force is an inequality, not an equation. We have

$$f \le \mu_s n \tag{11.10a}$$

where f is the distributed friction force along the bar and $n = mg/l$ is the uniform distributed normal force. For the bar to rotate around point O, the tendency of slip will be the highest away from point O and the slip direction would be reversed on one side versus the other. In this case, the bar will have a clockwise rotation tendency and the distributed friction force direction will be as shown in (b). However, we do not know the size of the distributed friction force because, according to Equation (11.10a), its size could be from 0 up to $\mu_s n$. Let us assume a general case of the distributed frictional force as shown in (b), in which a portion is saturated at $\mu_s n$ and the non-saturated portion is distributed linearly. With this assumption, we go through the ABCC procedure to complete the FBD as shown. Basically, we have an applied force (moment M), no body force in the direction of concern, one contact force, R, at the pin, and the contact distributed frictional force. From here, we only need one equilibrium equation with respect to point O for the answer. To do that, we convert the distributed frictional forces into concentrated forces at equivalent locations as shown in (c).

$$\left[\sum M\right]_0 = 0 \Rightarrow F_1 \frac{l-c}{2} + F_2 \frac{2}{3}b + F_2 \frac{2}{3}b + F_3 \left(b + \frac{1}{2}c\right) - M = 0 \tag{11.10b}$$

where $F_1 = \mu_s n(l - c - 2b)$, $F_2 = \mu_s n(\frac{1}{2}b)$, $F_3 = \mu_s n c$, and $a = b + c$.

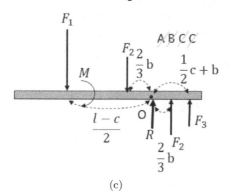

(c)

Solving Equation (11.10b), we have

$$M = \mu_s mgl \left(\frac{1}{2} - \frac{a}{l} + \frac{a^2}{l^2} - \frac{1}{3}\frac{b^2}{l^2} \right) \qquad (11.10\text{c})$$

When $b = 0$, the frictional force will reach its maximum and can no longer resist the moment to keep the bar from rotating. In other words, the moment needed to make the bar be on the verge of rotating will be

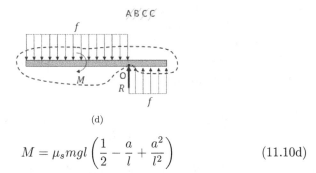

(d)

$$M = \mu_s mgl \left(\frac{1}{2} - \frac{a}{l} + \frac{a^2}{l^2} \right) \qquad (11.10\text{d})$$

The FBD for the bar at the verge of rotating will be as shown in (d). Finally, if we let $\frac{dM}{da} = 0$, we can determine the location of pin O which renders the moment M at its minimum. It is found that $M_{\min} = \frac{1}{4}\mu_s mgl$ and $a = \frac{l}{2}$.

Problem 11.11: A broom is held by two guy wires as shown in (a). There is a vertical load, P, acting at the tip of the boom. The weight of the boom is negligible compared with P. Calculate the minimum coefficient of static friction μ_s required between the boom and the ground so that the end of the boom will not slip.

(a)

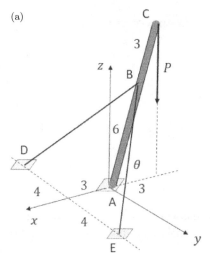

Solution: This is a fairly simple problem with calculation complexity due to the 3D configuration. The idea is to find out the normal contact force and the frictional force at point A. From them, the static friction coefficient can be determined. As pointed out before, for 3D problems, it is better to use math to solve the problem without attempting to draw elaborated 3D graphics for solutions.

To do so, we will define the coordinates of all relevant points in (a), as shown in (b). The tilting angle of the boom is determined to be $\theta = \cos^{-1} \frac{3}{9} = 70.53°$. The corresponding FBD is also shown in (b). We have an applied force, P, no body forces, three contact forces, F_x, F_y, and N, and two constraint forces, R_1 and R_2, along the guy wire directions. We then define these forces in the vector form as

$$\vec{R_1} = R_1 \frac{\vec{BD}}{|\vec{BD}|} = \frac{R_1}{|\vec{BD}|}(5\vec{i} - 4\vec{j} - 6\sin\theta\vec{k}) \qquad (11.11a)$$

$$\vec{R_2} = R_2 \frac{\vec{BE}}{|\vec{BE}|} = \frac{R_2}{|\vec{BE}|}(5\vec{i} + 4\vec{j} - 6\sin\theta\vec{k}) \qquad (11.11b)$$

$$\vec{P} = -P\vec{k}; \ \vec{F_x} = -F_x\vec{i}; \ \vec{F_y} = -F_y\vec{j}; \ \vec{N} = N\vec{k} \qquad (11.11c)$$

where $|\vec{BD}| = |\vec{BE}| = 8.544$.

(b)

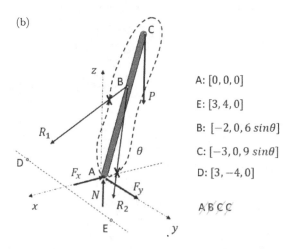

A: $[0, 0, 0]$

E: $[3, 4, 0]$

B: $[-2, 0, 6\sin\theta]$

C: $[-3, 0, 9\sin\theta]$

D: $[3, -4, 0]$

A B C C

We have five unknowns, R_1, R_2, F_x, F_y, and N. We will need five equations while we have six equilibrium equations. Therefore, the problem is solvable. At equilibrium, we have $\sum \vec{F} = 0$ and $\sum \vec{M} = 0$, which lead to

$$\left[\sum F\right]_x = 0 \Rightarrow \frac{5}{8.544} R_1 + \frac{5}{8.544} R_2 - F_x = 0 \qquad (11.11\text{d})$$

$$\left[\sum F\right]_y = 0 \Rightarrow \frac{-4}{8.544} R_1 + \frac{4}{8.544} R_2 - F_y = 0 \qquad (11.11\text{e})$$

$$\left[\sum F\right]_y = 0 \Rightarrow -P - \frac{5.657}{8.544} R_1 - \frac{5.657}{8.544} R_2 + N = 0 \qquad (11.11\text{f})$$

$$\left(\sum \vec{M}\right)_A = 0 \Rightarrow \vec{AB} \times \vec{R_1} + \vec{AB} \times \vec{R_2} + \vec{AC} \times \vec{P} = 0 \qquad (11.11\text{g})$$

where $\vec{AB} \times \vec{R_1} = \frac{R_1}{8.544} \begin{vmatrix} \vec{i} & \vec{j} & \vec{k} \\ -2 & 0 & 5.657 \\ 5 & -4 & -5.657 \end{vmatrix}$

$$= \frac{R_1}{8.544}(8\vec{k} + 5 \cdot 5.657\vec{j} + 4 \cdot 5.657\vec{i} - 2 \cdot 5.657\vec{j})$$

$$\vec{AB} \times \vec{R_2} = \frac{R_2}{8.544} \begin{vmatrix} \vec{i} & \vec{j} & \vec{k} \\ -2 & 0 & 5.657 \\ 5 & 4 & -5.657 \end{vmatrix}$$

$$= \frac{R_2}{8.544}(-8\vec{k} + 5 \cdot 5.657\vec{j} - 4 \cdot 5.657\vec{i} - 2 \cdot 5.657\vec{j})$$

$$\vec{AC} \times \vec{P} = -3P\vec{j}$$

From $[\sum M]_{Ax} = 0$, we have $R_1 = R_2$. From $[\sum M]_{Ay} = 0$, we have $R_1 = 0.755\,\text{P}$. From (11.11f), we have $N = 2\,\text{P}$. From (11.11d) and (11.11e), we have $F_y = 0$ and $F_x = 0.884\,\text{P}$. Finally, we have

$$\mu_s = \frac{F_x}{N} = 0.442 \qquad (11.11\text{h})$$

Problem 11.12: A submarine laboratory of the shell structure is submerged to the bottom of a shallow sea. Sea water enters the lower portion of the laboratory and is maintained at a level of 2 m by compressed air in the upper portion of the structure. The shell structure has a mass of 70,000 kg. At the bottom of the structure, a lead blast is attached. The density of the sea water is $\rho_1 = 1030\,\text{kg/m}^3$. The density of the lead is $\rho_2 = 11,340\,\text{kg/m}^3$. The air density at the atmosphere pressure of 101 kPa is $\rho_0 = 1.206\,\text{kg/m}^3$. Determine the gauge air pressure inside the laboratory and the mass of the lead blast so that the normal force between the submarine leg and the sea floor is 100 kN.

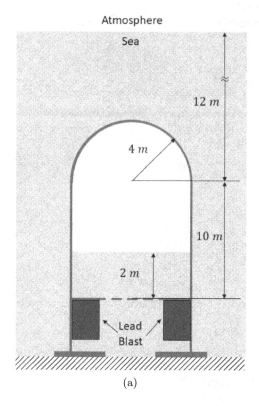

(a)

Solution: To solve this problem, we should know the principle of buoyance, which states that the buoyance force is equal to the weight of displaced portion of the fluid. Note that the displaced sea water in this problem includes the space of the laboratory structure and the space occupied by the lead blast. Second, we need to know the static pressure of sea water as a function of the depth, which is

$$p = p_0 + \rho g z \tag{11.12a}$$

Now, we can construct an FBD as shown in (b). Here, we do not have Applied forces; we have Body forces, $m_0 g$, $m_1 g$, and $m_2 g$, related to air, structure, and lead blast, respectively, two contact forces, the buoyance, F_B, and N, and finally, no constraint forces.

The buoyance is

$$F_B = \rho_1 (V_s + V_l) g \tag{11.12b}$$

where the structure volume $V_s = 8 \cdot \pi \cdot 4^2 + \frac{1}{2}(\frac{4}{3}\pi \cdot 4^3) = 536.1\,(m^3)$ and the volume of the lead blast $V_l = \frac{m_2}{\rho_2}$. The force equilibrium equation is

$$F_B - m_0 g - m_1 g - m_2 g + N = 0 \qquad (11.12c)$$

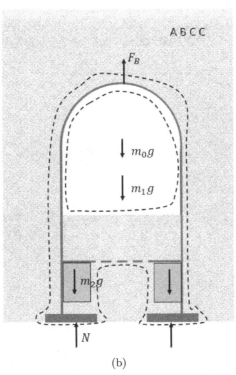

(b)

From (11.12a), we can determine the air pressure inside the laboratory as

$$p = p_0 + \rho_1 g z = p_0 + 1.03 \cdot 9.81 \cdot 20 = p_0 + 202\,(kPa) \qquad (11.12d)$$

Therefore, the gauge air pressure is 202 kPa and the absolute pressure is 303 kPa. The air density becomes three times based on the ideal gas law, i.e. $\hat{\rho}_0 = 3.618\,\mathrm{kg/m^3}$. Based on $\hat{\rho}_0$, $m_0 g = 3.618 \cdot 536.1 \cdot 9.81 = 19027\,(N)$. With Equation (11.12b) and $m_0 g$, Equation (11.12c) becomes

$$\rho_1\left(536.1 + \frac{m_2}{\rho_2}\right) 9.81 - 19027 - 70000 \cdot 9.81 - m_2 \cdot 9.81 + 100000 = 0$$

$$(11.12e)$$

Solving (11.12e), we obtain

$$m_2 = 539,415\,\text{kg} \qquad (11.12\text{f})$$

Problem 11.13: A system is configured as shown in (a). The horizontal member, ABC, is supported by four flexible cables but does not touch members DF and EF. Determine the force acting on members DF and EF.

(a)

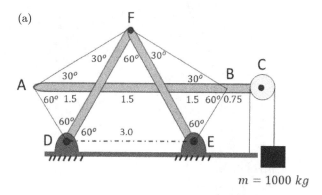

$m = 1000\ kg$

Solution: To solve this problem, we should first recognize that member ABC is not a truss member, while members DF and EF are. For the truss system (as discussed in problem 11.6), one approach is to trace one pin joint at a time, but the unknowns are limited to two. A larger FBD diagram can be constructed, but the unknowns are limited to three. The first approach is to find out the forces in the foundation joints as shown in (b). For this FBD, we have four unknowns. We will first see if we can solve for them partially. We have three equilibrium equations,

$$\left[\sum F\right]_x = 0 \Rightarrow D_x - E_x = 0 \qquad (11.13\text{a})$$

$$\left[\sum F\right]_y = 0 \Rightarrow -D_y + E_y - 2T = 0 \qquad (11.13\text{b})$$

$$\left[\sum M\right]_E = 0 \Rightarrow 3D_y - 1.5(2T) = 0 \qquad (11.13\text{c})$$

From (11.13c), we have $D_y = T$, pointing down. From (11.13b), we have $E_y = 3T$, where $T = 9.81\,\text{kN}$.

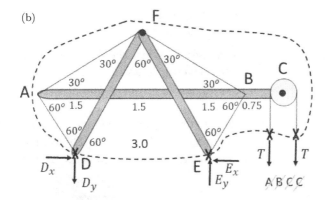

From here, we are pretty much stuck unless we can determine some forces from observation. Note that the cables are flexible; therefore, they can only sustain tension and no compression. Now examine member ABC, under the loading at C and restrained by four cables. With the loading as shown, member ABC can only rotate as shown in (c).

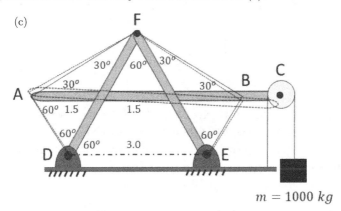

$$m = 1000 \ kg$$

From (c), apparently, cable BE can only be shortened because cables cannot sustain any compression. Therefore, the tension of cable BE should be zero. With this observation, an FBD for pin E will have only two unknowns, E_x and EF, which can be easily solved to obtain $E_x = 17\,\text{kN}$ and $EF = 34\,\text{kN}$ in compression. An FBD of pin D will now give $DF = 11.33\,\text{kN}$ in compression and tension of cable $AD = 22.66\,\text{kN}$.

If one does not observe that cable BE has zero force, then the other alternative is to construct all possible equations. We have eight unknowns, D_x, E_x, DF, EF, AD, BE, AF, and BF; therefore, we will need eight equations. We already have one as (11.13a). We need seven equations. We can get them with FBDs for pin D, pin E, and member ABC. However, solving eight equations is a formidable task.

Problem 11.14: A spring is used to adjust the belt tension for a motor to drive a device as shown in (a). The belt is a V-belt with geometry as shown. The motor transmits a constant-speed torque of 18 Nm when the belt is on the verge of slipping. Determine the spring force required to drive in (i) clockwise rotation and (ii) counter-clockwise rotation. The static friction coefficient between the pulley and the belt is 0.3. The combined mass of the motor assembly and the bracket is 35 kg at point G as shown.

(a)

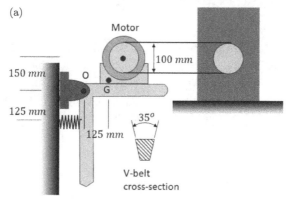

Solution: As always, we will start with the construction of a suitable FBD. Because no information was given about the driven device, we will not include it in the FBD. The FBD should also cut through the spring so that the spring force will appear in the governing equations. A suitable FBD is shown in (b).

(b)

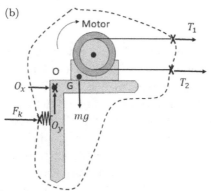

From (b), we have five unknowns. Therefore, we will need five equations. As a 2D problem, we have three equilibrium equations. The additional equations will be related to the belt tension as the transmitted torque is given and the belt is on the verge of slipping.

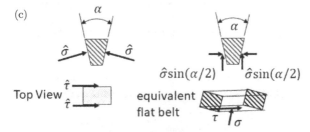

First, let us invoke the belt tension formulas (Equation (4.6) in Chapter 4). We have

$$dT = \tau R d\theta \quad \text{and} \quad T = \sigma R \tag{11.14a}$$

However, these equations are derived based on a flat belt as shown in Fig. 4.10. We need to convert the V-belt into an equivalent flat belt. This is done in (c). We have

$$\hat{\tau} = \mu_s \hat{\sigma} \tag{11.14b}$$

$$\sigma = 2\hat{\sigma} \sin\left(\frac{\alpha}{2}\right) \tag{11.14c}$$

$$\tau = 2\hat{\tau} \tag{11.14d}$$

Substituting Equations (11.14b) into (11.14a), we have

$$dT/T = \mu_s d\theta / \sin\left(\frac{\alpha}{2}\right) \tag{11.14e}$$

$$T_2 = T_1 e^{\mu_s \hat{\theta}} = T_1 e^{\mu_s \theta / \sin\left(\frac{\alpha}{2}\right)} \tag{11.14f}$$

For $\alpha = 35°$ and $\theta = \pi$, we have

$$T_2 = T_1 e^{0.3 \cdot \pi / \sin\left(\frac{\alpha}{2}\right)} = 23.14\, T_1 \tag{11.14g}$$

From the transmitted torque information, we also have, with respect to the center of the motor,

$$0.5\, T_2 - 0.5\, T_1 = 18 \tag{11.14h}$$

Now the moment equilibrium equation with respect to point O gives

$$0.125\, F_k - 0.1\, T_2 - 0.2\, T_1 - 0.125\, mg = 0 \tag{11.14i}$$

Solving Equations (11.14g–11.14i), we found $F_k = 671\,\text{kN}$ for clockwise rotation as shown in (a). For counter-clockwise rotation, the positions of

T_1 and T_2 will be reversed and Equation (11.14f) becomes,

$$0.125\,F_k - 0.2\,T_2 - 0.1\,T_1 - 0.125\,mg = 0 \qquad (11.14\mathrm{j})$$

which leads to $F_k = 959\,\mathrm{kN}$ for counter-clockwise rotation.

Problem 11.15: A gate (AB) is used to control the release of water in a fresh water reservoir as shown in (a). A counter-weight D is used to resist the water pressure to seal the opening. Determine the required mass and the force acting on member DB.

(a)

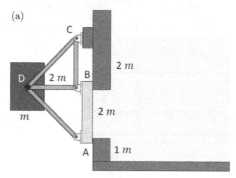

Solution: If we construct an FBD as shown in (b), we can use a moment equilibrium equation with respect to pin C for the solution for the mass. We do not have applied forces. We have body forces of the counter-weight. The contact forces include the force due to the contact static water pressure and the contact forces at the edge of the plate AB. Finally, we have constraint forces at pin C.

There are too many unknowns. For the design purposes, we will assume that the contact forces, F_A and F_B, are zero. With this assumption, the counter-weight will be the minimum. With this assumption, there are only three unknowns. We will first determine the total force due to the water pressure and the centroid of the pressure force. Based on the static hydrostatic pressure, we have

$$p = p_0 + \rho g z \qquad (11.15\mathrm{a})$$

Because the other side of the plate is air, we only need to use the gauge pressure. We have

$$p_B = \rho g z = 1000 \cdot 9.81 \cdot 2 = 19620\,(\mathrm{pa}) \qquad (11.15\mathrm{b})$$

$$p_A = \rho g z = 1000 \cdot 9.81 \cdot 4 = 39240\,(\mathrm{pa}) \qquad (11.15\mathrm{c})$$

The total hydrostatic force is found to be $F_w = (19620 + 39240) * \frac{2}{2} * 1.5 = 88290\,(N)$ and the centroid is found to be 1.111 m from the top edge of the plate (leave it to the student to work this out). Now, the moment equilibrium equation with respect to C becomes

$$2\,mg - 88290 \cdot (2 + 1.111) = 0 \tag{11.15d}$$

We easily obtain $m = 14000\,\text{kg}$, which is the minimal counter-weight needed. For design purposes, we should at least have 50% more weight to achieve a proper seal.

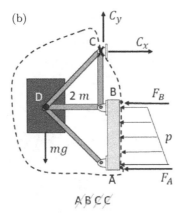

(b)

We can go ahead to determine the forces at pin C at the minimum counter-weight, which are $C_x = 88290\,\text{N}$ and $C_y = 137335\,\text{N}$. From here, the problem is simply a truss system problem. We can go from pin C to determine the force of member CB and CD. Then for pin D, we can find the force at member DB, which is 39245 N.

Problem 11.16: Determine the coordinates of the centroid of the volume shown in (a), which is obtained by revolving the right triangle through 90-degree about the z-axis.

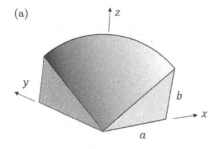

(a)

Solution: The general equations for determining the centroid of a 3D object are well known as

$$\bar{x} = \frac{\int xdV}{V}; \quad \bar{y} = \frac{\int ydV}{V}; \quad \bar{z} = \frac{\int zdV}{V} \tag{11.16a}$$

The key is always related to how to define the infinitesimal volume and to carry out the integration accordingly. A more useful form of Equation (11.16a) is

$$\bar{x} = \frac{\iint xdA_{yz}dx}{V}; \quad \bar{y} = \frac{\iint ydA_{xz}dy}{V}; \quad \bar{z} = \frac{\iint zdA_{xy}dz}{V} \tag{11.16b}$$

where $V = \int dA_{yz}dx = \int dA_{xz}dy = \int dA_{xy}dz$. A_{yz}, A_{xz}, and A_{xy} represent a surface parallel to the yz-plane, xz-plane, and xy-plane, respectively.

Sometimes, it might be possible to define an entire cross-section area without integration, such that

$$\bar{x} = \frac{\int xA_{yz}dx}{V}; \quad \bar{y} = \frac{\int yA_{xz}dy}{V}; \quad \bar{z} = \frac{\int zA_{xy}dz}{V} \tag{11.16c}$$

where $V = \int A_{yz}dx = \int A_{xz}dy = \int A_{xy}dz$.

The choice of surface is better if it is either symmetric or flat.

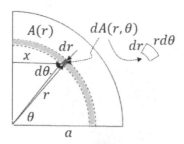

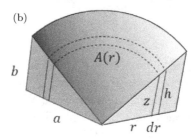

For this problem, we first consider how to determine $\bar{x}$ based on (11.16a). The volume of choice is shown in (b), as a cylindrical surface, $A(r)$, multiplying its thickness. With this definition, we can easily determine the total volume of the object as

$$V = \int_0^a A(r)\,dr = \int_0^a \frac{\pi}{2} rh(r)\,dr = \frac{\pi}{2}\frac{b}{a}\int_0^a r^2\,dr = \frac{\pi}{6}\frac{a^2}{b} \qquad (11.16d)$$

where $h(r) = \frac{b}{a}r$.

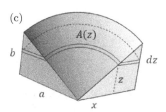

(c)

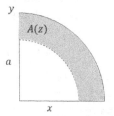

However, the surface, $A(r)$, has a different distance to the y-axis as shown in (b). Therefore, we need to break down the surface to infinitesimal sizes to define $dA(r,\theta) = r\,d\theta\,dr$, as shown. With this definition, we have $dV = h(r)\,dA(r,\theta)$, and

$$\bar{x} = \frac{\iint x\,h(r)\,dA(r,\theta)}{V} = \frac{\iint x\left(\frac{b}{a}\right) r \cdot r\,d\theta\,dr}{V}$$

$$= \frac{\int_0^a r\cos\theta\left(\frac{b}{a}\right) r \cdot r\,d\theta\,dr}{V} = \frac{3a}{2\pi} \qquad (11.16e)$$

Due to the symmetric feature of the object, $\bar{x} = \bar{y} = \frac{3a}{2\pi}$.

Now, we can determine $\bar{z}$ with the surface definition as shown in (c) based on Equation (11.16c). We choose a flat surface parallel to the xy-plane. The area of this flat surface can be easily defined as

$$A(z) = A_{xy} = \frac{\pi}{4}(a^2 - x^2) = \frac{\pi}{4}\left(a^2 - \frac{a^2}{b^2}z^2\right) \qquad (11.16f)$$

where $x = \frac{a}{b}Z$.

We can calculate the total volume of the object based on this surface of choice to verify the result of (11.16d).

$$V = \int_0^b A_{xy}dz = \int_0^b \frac{\pi}{4}\left(a^2 - \frac{a^2}{b^2}z^2\right)dz = \frac{\pi}{6}\frac{a^2}{b} \qquad (11.16g)$$

Because surface A_{xy} is parallel to the xy-plane, the z distance is the same for the entire surface. Based on Equation (11.16c), we have

$$\bar{z} = \frac{\int z A_{xy}dz}{V} = \frac{\int z\frac{\pi}{4}\left(a^2 - \frac{a^2}{b^2}z^2\right)dz}{V} = \frac{3}{8}b \qquad (11.16h)$$

The solution for $\bar{z}$ is easily obtained because a flat surface is chosen based on (11.16c).

For illustration purposes, we will examine different ways of solving for $\bar{x}$ based Equation (11.16c). To do so, we have to define a flat surface, A_{yz}, as shown in (d).

Note that the top edge of this surface, A_{yz}, is not a straight line.

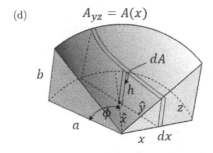

(d) $A_{yz} = A(x)$

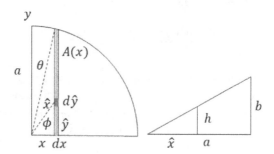

The task of determining the area of A_{yz} is not trivial. We need to determine the height, $h(\hat{x})$. Along A_{yz}, we can define angle ϕ, which is zero for the plane at the x-axis and $\phi = \theta$ at the y-axis. The height along A_{yz}

can be determined based on the triangle shown in (d) as

$$h = \frac{b}{a}\hat{x} \tag{11.16i}$$

Also from geometry, we have

$$x = \hat{x}\sin\phi; \quad \hat{y} = \hat{x}\cos\phi; \quad x = a\sin\theta \tag{11.16j}$$

and $\hat{y}$ goes from zero to $a\cos\theta$.

We define an infinitesimal area, $dA = hd\hat{y} = \frac{b}{a}\hat{x}d\hat{y}$, and we can integrate it to determine the area of $A_{yz}(x)$,

$$A_{yz} = \int dA = \int \frac{b}{a}\hat{x}d\hat{y} = \int \frac{b}{a}\frac{x}{\sin\phi}d\left(x\frac{\cos\phi}{\sin\phi}\right) = \int \frac{b}{a}\frac{x^2}{\sin\phi}d(\cot\phi)$$

$$= -\frac{b}{a}x^2 \int_{\pi/2}^{\theta} \frac{1}{(\sin\phi)^3}d\phi \tag{11.16k}$$

The above integration is not simple, and we simply look up an integration table to complete it. We have

$$A_{yz} = -\frac{b}{a}x^2 \int_{\frac{\pi}{2}}^{\theta} \frac{1}{(\sin\phi)^3} d\phi = \frac{b}{a}x^2 \left\{ \left[\frac{1}{2}\frac{\cos\phi}{(\sin\phi)^2}\right]\Big|_{\frac{\pi}{2}}^{\theta} \right.$$

$$\left. + \frac{1}{2}\ln|\csc\phi - \cot\phi|_{\frac{\pi}{2}}^{\theta} \right\}$$

$$= \frac{bx^2}{2a}\frac{\cos\theta}{(\sin\theta)^2} + \frac{bx^2}{2a}\left[\ln\left(\frac{1}{\sin\theta} + \frac{\cos\theta}{\sin\theta}\right)\right]$$

$$= \frac{ab}{2}\cos\theta + \frac{ab}{2}(\sin\theta)^2 \ln\left(\frac{\cos\theta + \sin\theta}{\sin\theta}\right) \tag{11.16l}$$

We can verify if Equation (11.16l) is correct. At $\theta = 0$, the area is that of the original triangle, and Equation (11.16l) gives a correct answer of $\frac{1}{2}a^2b$. At $\theta = \pi/2$, it is zero.

We can again calculate the volume of the object for verification purposes. We have

$$V = \int_0^a A_{yz}dx = \int_0^a \frac{ab}{2}\cos\theta dx + \int_0^a \frac{ab}{2}(\sin\theta)^2 \ln\left(\frac{\cos\theta + \sin\theta}{\sin\theta}\right) dx$$

$$= \frac{a^2b}{2}\int_0^{\frac{\pi}{2}} (\cos\theta)^2 d\theta + \frac{a^2b}{2}\int_0^{\frac{\pi}{2}} (\sin\theta)^2 \ln\left(\frac{\cos\theta + \sin\theta}{\sin\theta}\right)\cos\theta\, d\theta$$

$$= \frac{\pi}{8}a^2b + \frac{\pi}{24}a^2b = \frac{\pi}{6}a^2b \tag{11.16m}$$

The integration of the above equation, in particular the second term, is not trivial. For the second term, integration by part is used to complete the integration. This will be left for readers to go through as a fun project. A hint for the integration of the second term is as follows:

$$\frac{a^2 b}{2} \int_0^{\frac{\pi}{2}} (\sin \theta)^2 \ln \left(\frac{\cos \theta + \sin \theta}{\sin \theta} \right) \cos \theta \, d\theta$$

$$= \frac{a^2 b}{2} \frac{1}{3} \int \ln \left(\frac{\cos \theta + \sin \theta}{\sin \theta} \right) d (\sin \theta)^3 \qquad (11.16n)$$

From here, we can try to determine $\bar{x} = \frac{\iint x A_{yz} \, dx}{V}$. Judging from the complexity of the integration involved in (11.16m), this integration will be even more complicated. Again, this will be left for inspired readers to go through.

(e)   (f)   (g)

Finally, we can consider a way to convert the object as the combination of different objects to determine the centroid. As shown in (e–g), a circular disk is the combination of a cone (f) and an object with a cone cavity (e). If we take (e) and divide it to four equal parts, we have (a) of the original problem. Or we can take out a quarter cone off a quarter pie (h) to reach the object of (a).

(h)

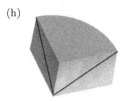

We can invoke the equation

$$\bar{x} V = \bar{x}_1 V_1 + \bar{x}_2 V_2 + \bar{x}_3 V_3 + \cdots \qquad (11.16o)$$

$$\bar{y} V = \bar{y}_1 V_1 + \bar{y}_2 V_2 + \bar{y}_3 V_3 + \cdots \qquad (11.16p)$$

$$\bar{z} V = \bar{z}_1 V_1 + \bar{z}_2 V_2 + \bar{z}_3 V_3 + \cdots \qquad (11.16q)$$

In this case, we have a quarter disk (h) whose centroid can be easily found from a table. Similarly, the centroid of a quarter cone can be either found from a table or by integration. Finally, we have

$$\bar{x}_{\text{object}} = (\bar{x}_{\text{disk}} V_{\text{disk}} - \bar{x}_{\text{cone}} V_{\text{cone}}) / V_{\text{object}} \qquad (11.16\text{r})$$

$$\bar{y}_{\text{object}} = (\bar{y}_{\text{disk}} V_{\text{disk}} - \bar{y}_{\text{cone}} V_{\text{cone}}) / V_{\text{object}} \qquad (11.16\text{s})$$

$$\bar{z}_{\text{object}} = (\bar{z}_{\text{disk}} V_{\text{disk}} - \bar{z}_{\text{cone}} V_{\text{cone}}) / V_{\text{object}} \qquad (11.16\text{t})$$

Again, this will be left for the reader to practice.

Problem 11.17: A semicircular beam was fixed on one end and suspended as a curved cantilever beam, as shown in (a) on a vertical plane. The total mass of the beam is m. Determine the bending moment in the beam as a function of θ.

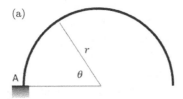

(a)

Solution: The solution is quite straightforward. We will first construct an FBD for the entire beam as shown in (b). We have no applied force, one body force at the mass center G, no contact forces, and one constraint at the left side edge of the beam. This constraint is a total constraint; therefore, it contains forces and moments. We easily determine that the force and the moment are as shown in (b).

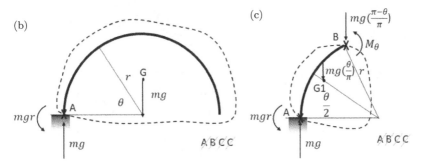

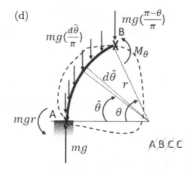

(d)

To determine the bending moment in the beam at a specific angle, we construct another FBD as shown in (c). By looking up the centroid table, we know that the centroid locates at G_1 as shown in (c) and the radial distance, $r_{G_1} = \frac{r \sin(\frac{\theta}{2})}{(\frac{\theta}{2})}$.

For the FBD of (c), we have applied forces as the constraint forces from (b) at the left edge, one body force at G_1, no contact forces, and one constraint force at the right edge where the boundary cuts through the beam. Similarly, this is a full constraint. We have two equilibrium equations to determine the constraint force and moment at the right edge. The vertical force is easily determined as shown in (c). The moment equation is

$$\left[\sum \vec{M}\right]_A = 0 \Rightarrow mgr - mg\left(\frac{\theta}{\pi}\right)\left(r - r_{G_1}\cos\frac{\theta}{2}\right)$$

$$- mg\left(\frac{\pi - \theta}{\pi}\right)(r - r\cos\theta) + M_\theta = 0 \qquad (11.17a)$$

From Equation (11.17a), we have $M_\theta = -\frac{mgr}{\pi}((\pi - \theta)\cos\theta + \sin\theta)$. As it is negative, the actual direction of M_θ should be CW, not CCW as shown in (c).

If we do not have the centroid table to look up r_{G_1}, then we can solve the problem by considering the distributed gravitational force at the partial beam as shown in (d).

The moment equilibrium equation becomes

$$\left[\sum \vec{M}\right]_A = 0 \Rightarrow mgr - \int_0^\theta (r - r\cos\hat{\theta})mg\left(\frac{d\hat{\theta}}{\pi}\right)$$

$$- mg\left(\frac{\pi - \theta}{\pi}\right)(r - r\cos\theta) + M_\theta = 0 \qquad (11.17b)$$

We will obtain the same answer by solving Equation (11.17b).

Problem 11.18: A wedge is inserted into a split ring with a force P as shown in (a). The friction coefficient between the wedge and the ring is μ_s. The force P is then removed. Due to self-locking, the wedge stays in place to split the ring. Determine the maximum residual bending moment in the ring.

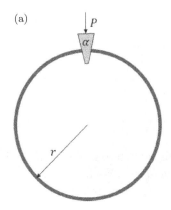

(a)

Solution: We first consider the condition that the wedge has gone as much as it can go with the force P. This force cannot drive the wedge further because of the contact forces between the ring and the wedge, which include normal and frictional forces. We construct an FBD as shown in (b) under such a condition. We have an applied force, no body force, contact forces, and no constraint forces. The normal force and the frictional force under such a condition can be related as

$$F_{f1} = \mu_s N \tag{11.18a}$$

The force equilibrium equation along the vertical direction is

$$2N \sin \frac{\alpha}{2} - P + 2F_{f1} \cos \frac{\alpha}{2} = 0 \tag{11.18b}$$

Solving Equations (11.18a) and (11.18b), we have

$$N = \frac{P}{2 \left(\sin \frac{\alpha}{2} + \mu_s \cos \frac{\alpha}{2} \right)} \tag{11.18c}$$

Note that this force N is due to the elastic deformation of the ring. After force P is removed, force N stays the same while the frictional force will change as long as it is within the static friction limit.

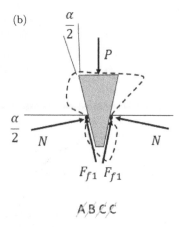

(b)

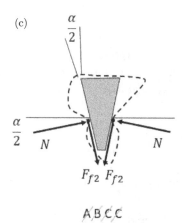

(c)

If the friction is high enough to achieve self-locking, the wedge will stay in place after force P is removed. We have a new force equilibrium condition as shown in (c). The new force equilibrium equation is

$$2N \sin \frac{\alpha}{2} - 2F_{f2} \cos \frac{\alpha}{2} = 0 \tag{11.18d}$$

We then have

$$F_{f2} = N \tan \frac{\alpha}{2} \tag{11.18e}$$

Both forces N and F_{f2} will contribute to the bending moment of the ring. The maximum bending moment will be at the bottom point. From here, we simply calculate the moment due to N and F_{f2} with respect to the bottom point. The solution is

$$M_{\max} = 2Pr/\left(\sin\alpha + 2\mu_s\left(\cos\frac{\alpha}{2}\right)^2\right) \qquad (11.18\text{f})$$

Problem 11.19: A paraboloidal dish is used as a focusing mirror for collecting starlight. The density of the dish is ρ per unit area. Determine the moment needed at point O so that the mirror can remain at its horizontal position as shown in (a).

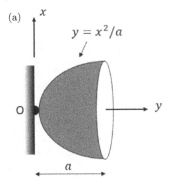

(a)

$y = x^2/a$

Solution: The first step for solving this problem is to draw an appropriate FBD as shown in (b). It is clear that the moment needed is $mg\bar{y}$, where $\bar{y}$ is the centroid distance from point O along the y-axis. We will need to determine mg and $\bar{y}$. To determine m, we need to know the total area of the disk. Because this disk is a paraboloidal curve revolving around the y-axis, we will need to conduct integration. Because of the steep curvature of the paraboloidal curve, the integration is rather complex in order to obtain correct answers.

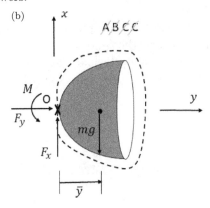

(b)

As shown in (c), an infinitesimal segment along the paraboloidal curve is $dL = \sqrt{1 + 4x^2/a^2}\, dx$. Readers should refer to Section 1.2.4 to ensure the error ratio of the defined dL is negligible. Revolving this small segment,

$$A = \int dA = \int_0^a 2\pi x \sqrt{1 + \frac{4x^2}{a^2}}\, dx = 2\pi \left.\left(\frac{a^2}{12}\left(1 + \frac{4x^2}{a^2}\right)^{\frac{3}{2}}\right)\right|_0^a = 5.3304a^2$$

(11.19a)

A great tool from WolframAlpha.com can be used to do the integration for convenience. This online tool can be accessed at wolframalpha.com[2]

$$(c) \qquad dL = \sqrt{1 + 4x^2/a^2}\, dx$$

The centroid $\bar{y}$ is determined as

$$\bar{y} = \frac{\int y\, dA}{\int dA} = \int_0^a 2\pi x y \sqrt{1 + \frac{4x^2}{a^2}}\, dx/A$$

$$= \frac{\left.\left(\frac{\pi}{60a}\right)(6x^2 - a^2)(a^2 + 4x^2)\sqrt{1 + \frac{4x^2}{a^2}}\right|_0^a}{A} = \frac{2.9793a^3}{5.3304a^2}$$

$$= 0.5589a$$

(11.19b)

Finally, the moment needed is found to be

$$M = \rho A g \bar{y} = 2.9793\rho g a^3$$

(11.19c)

Problem 11.20: Determine the principal moments of inertia (maximum and minimum) with respect to the centroid of the triangle area shown in (a) and identify the angle of the principal axes with respect to the $x - y$ coordinate. Use the Mohr circle of inertia to solve this problem.

---

[2]https://www.wolframalpha.com/input/?i=x2%2Fa.

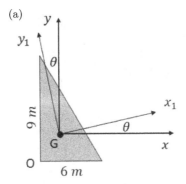

Solution: First, we readily find the moments of inertia with respect to the x-axis and y-axis using a moment of inertia table,

$$I_{xx} = \frac{6 \times 9^3}{36} = 121.5 \tag{11.20a}$$

$$I_{yy} = \frac{9 \times 6^3}{36} = 54 \tag{11.20b}$$

Next, we will determine the product of inertia with respect to the $x - y$ coordinate centered at point O, as shown in (b). For the shaded infinitesimal area shown in (b), we have

$$dA = ydx = 9\left(1 - \frac{x}{6}\right)dx \tag{11.20c}$$

The product of inertia of dA is defined as

$$dI_{xyo} = x_{el}y_{el}dA = x\left(\frac{1}{2}y\right)dA = \frac{1}{2}xy^2dx = \frac{1}{2}x(9^2)\left(1 - \frac{x}{6}\right)^2 dx \tag{11.20d}$$

$$I_{xyo} = \int dI_{xyo} = \int_0^6 \frac{1}{2}x(9^2)\left(1 - \frac{x}{6}\right)^2 dx = 121.5 \tag{11.20e}$$

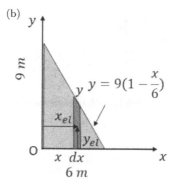

(b)

$y = 9(1 - \dfrac{x}{6})$

Now, we only need to shift the coordinate to point G to find the corresponding product of inertia.

$$I_{xyG} = I_{xyo} - \bar{x}\bar{y}A = 121.5 - 2 \times 3 \left(\frac{1}{2} \times 6 \times 9 \right) = -40.5 \qquad (11.20\text{f})$$

From here, we are ready to use Mohr's circle of inertia to find out the principal moments of inertia. Before doing so, we will provide a brief discussion of how to construct Mohr's circle of inertia. As shown in (c), the moments of inertia with respect to the $x_1 - y_1$ coordinate can be related to the ones of the $x - y$ coordinate as

$$I_{x_1 x_1} = \int y_1 y_1 dA = \int (y \cos\theta - x \sin\theta)^2 dA \qquad (11.20\text{g})$$

$$I_{y_1 y_1} = \int x_1 x_1 dA = \int (x \cos\theta + y \sin\theta)^2 dA \qquad (11.20\text{h})$$

$$I_{x_1 y_1} = \int x_1 y_1 dA = \int (x \cos\theta + y \sin\theta)(y \cos\theta - x \sin\theta) dA \quad (11.20\text{i})$$

$$I_{x_1 x_1} + I_{y_1 y_1} = I_{xx} + I_{yy} \qquad (11.20\text{j})$$

After some lengthy derivations, we can obtain the following relationship:

$$\left(I_{x_1 x_1} - \frac{I_{xx} + I_{yy}}{2} \right)^2 + (I_{x_1 y_1})^2 = \left(\frac{I_{xx} - I_{yy}}{2} \right)^2 + I_{xy}^2 \qquad (11.20\text{k})$$

and

$$I_{x_1x_1} = \frac{I_{xx} + I_{yy}}{2} + \frac{I_{xx} - I_{yy}}{2}\cos 2\theta - I_{xy}\sin 2\theta \qquad (11.20\text{m})$$

$$I_{y_1y_1} = \frac{I_{xx} + I_{yy}}{2} - \frac{I_{xx} - I_{yy}}{2}\cos 2\theta + I_{xy}\sin 2\theta \qquad (11.20\text{n})$$

$$I_{x_1y_1} = \frac{I_{xx} - I_{yy}}{2}\sin 2\theta + I_{xy}\cos 2\theta \qquad (11.20\text{o})$$

Equation (11.20k) can be constructed graphically, denoted as Mohr's circle of inertia, as shown in (d).

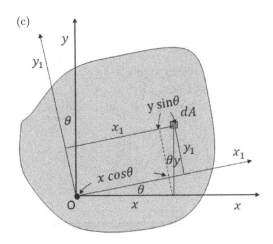

If we arrange the axis of $I_{x_1y_1}$ to point downward as positive, we can define a Mohr's circle as shown in (d) consistent with the angle rotation as in (c). It can then be proved that the moments of inertia with respect to the $x_1 - y_1$ coordinate locate at the points with a 2θ rotation on the circle, as shown in (d). From Mohr's circle, we can also find out after rotating an angle of 2α, we will find maximum and minimum moments of inertia as I_1 and I_2, respectively. The corresponding coordinates for I_1 and I_2 are denoted as the principal axes.

(d)

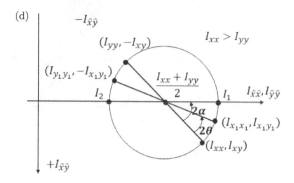

We can now draw the corresponding Mohr's circle for the triangle area of (a). The steps are quite simple. First, we determined the locations of (I_{xx}, I_{xy}) and (I_{yy}, I_{xy}) as $(121.5, -40.5)$ and $(54, 40.5)$ from Equations (11.20a), (11.20b), and (11.20f), and then we marked them accordingly on Mohr's circle, as shown in (e). We can then draw the Mohr circle which contains both points. The center of the circle is determined to be $\frac{121.5+54}{2} =$ 87.75. The radius of the circle is found to be $\sqrt{(121.5 - 87.75)^2 + 40.5^2} =$ 52.72. The values of I_1 and I_2 can then be readily determined to be 140.47 and 35.03, respectively. Finally, the principal axis is found to be clockwise rotation by

$$\alpha = \frac{1}{2} \tan^{-1} \frac{40.5}{121.5 - 87.75} = 25.1° \qquad (11.20p)$$

(e)

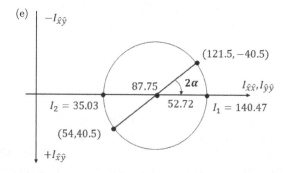

Problem 11.21: As shown in (a), a conical pivot bearing is used to support a shaft thrust, subjected to an axial force. Determine the moment needed to overcome the friction for turning the shaft under different contact conditions between the shaft and bearing. The friction coefficient is assumed to be μ. New bearing condition (1): the contact pressure between the shaft and the

bearing is uniform, as shown in (b). Worn bearing condition (2): the contact pressure is as shown in (c).

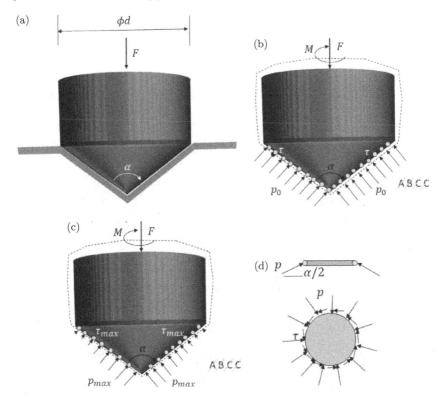

Solution: The FBDs for both conditions are shown in (b) and (c), respectively. For Condition (1), the distributed contact pressure is p_0 and the distributed friction is $\tau = \mu p_0$. We need to first define p_0 as a function of F. As shown in (d), we first slice the cone into thin layers. The vertical force acting on each thin layer is

$$dF = p_0 dA \sin\left(\frac{\alpha}{2}\right) = p_0 r d\theta \left(\frac{dr}{\sin\left(\frac{\alpha}{2}\right)}\right) \sin\left(\frac{\alpha}{2}\right) = p_0 r\, d\theta\, dr \quad (11.21a)$$

Integrating Equation (11.21a) over the entire cone surface, we have

$$F = \int_0^{2\pi} \int_0^{d/2} p_0 r\, dr\, d\theta = \left(\frac{\pi d^2}{4}\right) p_0 \quad (11.21b)$$

The uniform pressure is

$$p_0 = \frac{4F}{\pi d^2}. \tag{11.21c}$$

The corresponding distributed friction force is $\tau = \mu p_0$ and the corresponding frictional moment over a small cone surface area is

$$dM = (\tau dA)r = \int_0^{2\pi} \int_0^{d/2} \mu p_0 r \cdot r d\theta \left(\frac{dr}{\sin\left(\frac{\alpha}{2}\right)} \right) \tag{11.21d}$$

Integrating Equation (11.21d), we have

$$M_1 = \frac{\mu F d}{3 \sin\left(\frac{\alpha}{2}\right)} \tag{11.21e}$$

We repeat the same procedures above for Condition (2). The contact pressure distribution can be found as

$$p = \left(-\frac{2}{d}r + 1 \right) p_{\max} \tag{11.21f}$$

Replacing p_0 of Equation (11.21b) with p of Equation (11.21f), we have

$$p_{\max} = \frac{12F}{\pi d^2} \tag{11.21g}$$

Similarly, replacing p_0 of Equation (11.21d) with p, we have

$$M_2 = \frac{\mu F d}{4 \sin\left(\frac{\alpha}{2}\right)} \tag{11.21h}$$

Problem 11.22: Determine the force P needed to achieve equilibrium of the hinged masses as shown in (a). The mass center of each mass is in the center of its respective rectangle. Neglect the masses of the legs. If $b = a$, at which angle of θ will equilibrium occur with $P = 0$?

Solution: It is easier to solve this problem using the method of virtual work discussed in Section 9.4. To do so, we still need to construct an FBD first. However, it is obvious that we should construct an FBD for the entire system as shown in (b). There are three different types of force acting on the system. Among them, Mg is a conservative force, P is a non-conservative force that does work, and N is a non-conservative force that does not do work. Notice that all hinges are assumed to be frictionless, and there is no friction between the roller and the ground.

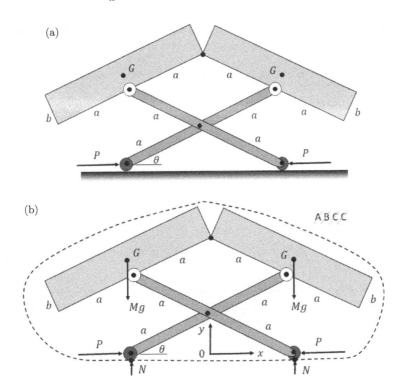

For Mg, we define the corresponding potential energy with respect to the datum of the $x - y$ coordinate defined in (b) as

$$U = 2Mg \left(2a \sin \theta + \frac{1}{2} b \cos \theta \right) \tag{11.22a}$$

The change of the potential energy is

$$\delta U = Mg(4a \cos \theta \delta\theta - b \sin \theta \delta\theta) \tag{11.22b}$$

The total virtual work done by both P forces when the rollers are pushed toward each other is

$$\delta W = 2P\delta x = 2P a \sin \theta \, \delta\theta \tag{11.22c}$$

Here we need to be a bit careful about the work being positive or negative. For the roller on the right, its position vector is $\vec{x} = a \cos \theta \vec{i}$, which gives a derivative of $\delta\vec{x} = -a \sin \theta \, \delta\theta \, \vec{i}$. The force vector is $\vec{P} = -P\vec{i}$. Therefore, the work is $\vec{P} \cdot \delta\vec{x} = Pa \sin \theta \, \delta\theta$. It is easy to see that the work done at the left roller is the same and is positive, leading to the total virtual work of Equation (11.20c). Now, applying the method of the virtual work, we have

$$\delta W - \delta U = 0 = 2Pa\sin\theta\,\delta\theta - Mg(4a\cos\theta\,\delta\theta - b\sin\theta\,\delta\theta) \qquad (11.22\text{d})$$

Solving Equation (11.22d), we obtain

$$P = Mg\left(2\cot\theta - \frac{b}{2a}\right) \qquad (11.22\text{e})$$

If $P = 0$ and $b = a$, Equation (11.22d) is reduced to

$$-\delta U = 0 = -Mg(4a\cos\theta\,\delta\theta - a\sin\theta\,\delta\theta) \qquad (11.22\text{f})$$

From the above equation, we obtain

$$\theta = 75.96° \qquad (11.22\text{g})$$

Problem 11.23: Determine the x-coordinate of the centroid of the shaded area shown in (a) and find the radius of gyration of the area about the vertical centroidal axis. The curved boundary is parabolic with zero slope at the y-axis.

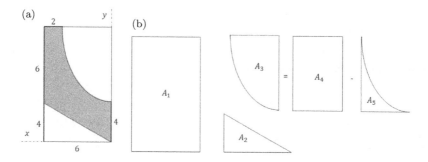

Solution: An effective way to solve the centroid problem of an area with an irregular shape is to represent the area with a combination of several areas with regular shapes listed in the table of centroids as shown in (b). The parabolic curve can be represented by $y = \frac{3}{8}x^2$.

We have

$$A = A_1 - A_2 - A_3 = A_1 - A_2 - (A_4 - A_5) = 32 \qquad (11.23\text{a})$$

where $A_1 = 60$, $A_2 = 12$, $A_3 = A_4 - A_5 = 24 - \int dA_5 = 16$, and $\int dA_5 = \int_0^4 y\,dx = \int_0^4 \frac{3}{8}x^2\,dx = 8$.

To calculate the centroid, we will calculate the first moment of the area as

$$\int x dA = \int x \, dA_1 - \int x \, dA_2 - \int x \, dA_3 = \int x \, dA_1$$

$$- \int x \, dA_2 - \left(\int x \, dA_4 - \int x \, dA_4 \right) = 108 \quad (11.23b)$$

where $\int x \, dA_1 = A_1 \cdot 3 = 180$

$$\int x \, dA_2 = A_2 \cdot 4 = 48$$

$$\int x \, dA_3 = \int x \, dA_4 - \int x \, dA_5 = A_4 \cdot 2 - \int_0^4 x \cdot \frac{3}{8} x^2 dx = 24$$

Therefore, $\bar{x} = \frac{108}{32} = 3.375$.

To determine the radius of gyration about the y-axis, we will first calculate the moment of inertia as

$$I_{yy} = \int x^2 dA = \int x^2 dA_1 - \int x^2 dA_2 - \int x^2 dA_3 = 452.8 \quad (11.23c)$$

where $\int x^2 dA_1 = \frac{1}{3} 10 \cdot 6^3 = 720$

$$\int x^2 \, dA_2 = \int_0^6 \frac{2}{3} x \cdot x^2 dx = 216$$

$$\int x^2 \, dA_3 = \int x^2 dA_4 - \int x^2 \, dA_5 = \frac{1}{3} 6 \cdot 4^3 - \int \frac{3}{8} x^2 \cdot x^2 dx = 51.2$$

Therefore,

$$I_{\bar{y}\bar{y}} = I_{yy} - A\bar{x}^2 = 452.8 - 365.6 = 88.3 \quad (11.23d)$$

Finally, the radius of gyration is $\bar{k} = \sqrt{I_{\bar{y}\bar{y}}/A} = 1.66$.

Problem 11.24: Construct the shear and moment diagrams for the loaded beam shown in (a). Indicate the maximum magnitude of the bending moment and the location.

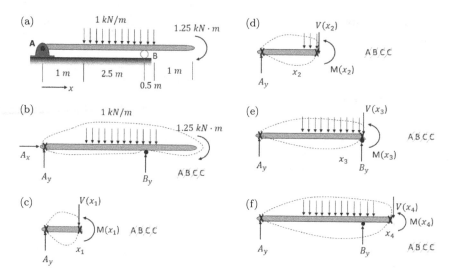

Solution: To construct the shear and moment diagrams, we need to construct a series of FBDs. The first one is for the overall system as shown in (b). From (b), we obtain $A_x = 0$, $A_y = 0.5\,\text{kN}$, and $B_y = 2.5\,\text{kN}$. We will define the downward shear force, $V(x)$, and counter-clockwise moment, $M(x)$, as positive when we construct the shear and moment diagram.

The subsequent FBDs are shown in (c)–(f). The FBD of (c) is applicable for $0 < x_1 < 1$. We found that $V(x_1) = 0.5$, downward (i.e. positive in the shear diagram), as shown in (g). From the moment equilibrium equation, we found that $M(x_1) = x_1 V(x_1) = 0.5x_1$. We plot it accordingly in (g).

The FBD of (d) is applicable for $0 < x_2 < 3.5$. We found that $V(x_2) = 1.5 - x_2$, which changes from 0.5 to -2, as shown in (g). From the moment equilibrium equation, we found that $M(x_2) = -0.5x_2^2 + 1.5x_2 - 0.5$. We found that $M(1.5) = 0.625$ and $M(3.5) = -1.375$. We need to plot this as a second-order polynomial curve, as shown in (g). An easier way, which is taught in the course of the strength of materials, is that we realize that the moment is the integration of the shear diagram. As a result, the moment diagram can be obtained graphically, which is much easier. We actually see that in $V(x_2)$ and $M(x_2)$ obtained above.

The FBD of (e) is applicable for $3.5 < x_3 < 4$. We found that $V(x_3)$ jumps from -2 to 0.5 because of B_y. We found that $V(x_3) = 4 - x_3$.

To determine the moment function, we will simply integrate the shear force function. We found that $M(x_3) = -0.5x_3^2 + 4x_3 - 9.25$. Because the shear force is not continuous at $x_3 = 3.5$, the slope of the moment curve at $x_3 = 3.5$ is not smooth. Immediately before $x_3 = 3.5$, the slope is -2, and immediately after it is 0.5.

The FBD of (f) is applicable for $4 < x_4 < 5$. We found that $V(x_4) = 0$, and therefore $M(x_4) = -1.25$, as a constant, until $x_4 = 5$, at which the moment jumps to zero due to the external moment applied at the end of the beam.

From (g), the largest magnitude of the moment is 1.375 kNm at $x = 3.5$ m.

Finally, with some practice, the shear and moment diagrams can be constructed graphically without finding out the exact functions and no need for constructing those additional FBDs of (c)–(f).

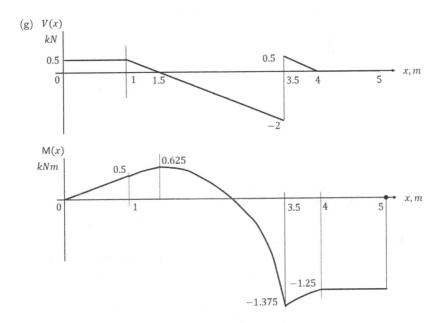

Problem 11.25: The cable hangs under its own weight and is suspended at two ends as shown in (a). Determine the tension of the cable at any point and the curvature of the cable. What is the total length of the cable?

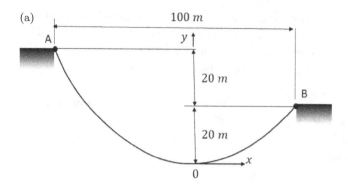

(a)

100 m

A

y ↑

20 m

B

20 m

0

x

Solution: We will first draw an FBD for the entire cable as shown in (b) and conduct some initial analysis.

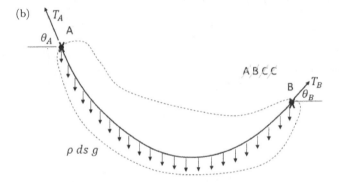

(b)

T_A

A

θ_A

$\rho\, ds\, g$

A B C C

B

T_B

θ_B

Following the ABCC procedure, we note that there are no applied forces (A). For the body force, we have distributed body force (B) along the cable as $\rho ds\, g$, where ρ is the density per unit length and ds is an infinitesimal length. There are no contact forces (C), but there are two constraint forces (C). These constraints are cable constraints, and we assume that the cable tension is along the tangential direction at point A and point B. To reach equilibrium, we have

$$T_A \cos\theta_A = T_B \cos\theta_B \qquad (11.25a)$$

$$T_A \sin\theta_A + T_B \sin\theta_B = \rho sg \qquad (11.25b)$$

where s is the entire cable length. There are five unknowns involved in the above two equations, and we have no way to solve for any of them.

We have to construct another FBD. One choice is shown in (c).

(c) A̶B̶C̶C̶

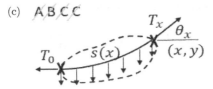

This FBD cuts at the lowest point where its tangential is horizontal. The other end of the boundary cuts at an arbitrary point of (x, y).

Now, it becomes clear that

$$T_A \cos\theta_A = T_B \cos\theta_B = T_0 = T_x \cos\theta_x \qquad (11.25c)$$

In other words, the horizontal component of the cable tension at any point is a constant.

From (c) and $s_x = s(x)$, we can write the force equation in the vertical direction as

$$T_x \sin\theta_x = \rho s_x g \qquad (11.25d)$$

Combining Equations (11.25c) and (11.25d), we have

$$\tan\theta_x = \rho s_x g / T_0 \qquad (11.25e)$$

From cable curvature, we also have

$$\frac{dy}{dx} = \tan\theta_x \qquad (11.25f)$$

Combining Equations (11.25e) and (11.25f), we have

$$\frac{dy}{dx} = \frac{\rho s_x g}{T_0} = \frac{s_x}{a} \qquad (11.25g)$$

where $a = \frac{T_0}{\rho g}$, which is a constant.

If we know the length as a function of x, then we can carry out the integration of Equation (11.25g). From geometry, we know that

$$ds = \sqrt{dx^2 + dy^2} = \sqrt{1 + \left(\frac{dy}{dx}\right)^2}\, dx \qquad (11.25h)$$

Introducing Equation (11.25g) into Equation (11.25h), we have

$$ds = \sqrt{dx^2 + dy^2} = \sqrt{1 + \left(\frac{s}{a}\right)^2}\, dx \qquad (11.25i)$$

Rewriting Equation (11.25i) in its integration form, we have

$$\int \frac{1}{\sqrt{1 + \left(\frac{s}{a}\right)^2}} ds = \int dx \qquad (11.25j)$$

From Equation (11.25j), we obtain

$$x = a \ln \left[\frac{s}{a} + \sqrt{\left(\frac{s}{a}\right)^2 + 1} \right] \qquad (11.25k)$$

If we reformulate Equations (11.25g) and (11.25i), we have

$$\int \frac{s}{\sqrt{a^2 + s^2}} ds = \int dy, \qquad (11.25m)$$

which leads to

$$y = \sqrt{a^2 + s^2} - a \qquad (11.25n)$$

From Equations (11.25k) and (11.25n), we can plot the curve of the cable as a function of s. This is the famous Catenary curve. The tangential angle can be obtained from Equation (11.25e), also as a function of s.

Note that the same equations are used for the curvature to the left and the right of the lowest point.

For the cable system of (a), we will need to use Equations (11.25e), (11.25k), and (11.25n) to determine the value of a.

We construct the following five equations based on the geometric information given in (a).

$$y_B = 20 = \sqrt{a^2 + s_B^2} - a \qquad (11.25o)$$

$$y_A = 40 = \sqrt{a^2 + s_A^2} - a \qquad (11.25p)$$

$$x_B = a \ln \left[\frac{s_B}{a} + \sqrt{\left(\frac{s_B}{a}\right)^2 + 1} \right] \qquad (11.25q)$$

$$x_A = a \ln \left[\frac{s_A}{a} + \sqrt{\left(\frac{s_A}{a}\right)^2 + 1} \right] \qquad (11.25r)$$

$$x_A + x_B = 100 \qquad (11.25s)$$

There are five unknowns, x_B, x_A, s_B, s_A, and a, in the above five equations. We can program a Matlab script to solve for them easily. Here is the Matlab script:

```
clear a x20 x40 s20 s40
syms a x20 x40 s20 s40
% a is the parameter ;
% x20 is x position when y=20
% x40 is the x position when y=40
% s20 is the s distance when y=20
% s40 is the s distance when y=40


eq1 = x20==a*log(s20/a + (s20^2/a^2 +1)^0.5);
eq2 = x40==a*log(s40/a + (s40^2/a^2 +1)^0.5);
eq3 = x20+x40==100;
eq4 = 20==(a^2+s20^2)^0.5 -a;
eq5 = 40==(a^2+s40^2)^0.5 -a;


eqns = [eq1,eq2, eq3, eq4, eq5 ];
S = solve(eqns);
sol = [S.a; S.x20; S.x40; S.s20; S.s40]
```

Executing this simple script, we obtain $a = 47.36$, $x_B = 43.12$, $x_A = 57.88$, $s_B = 47.90$, $s_A = 73.41$.

The total cable length is then found to be $s_{tatal} = 121.31\,m$. From the value of a, we determine the horizontal cable tension as $T_0 = 47.36\,\rho g$. The vertical tension component at B is $T_{By} = \rho s_B g = 47.90\,\rho g$. The total cable tension at B is $T_B = \sqrt{T_0^2 + T_{By}^2} = 67.36\rho g$, and the angle at B is $\theta_B = 45.32°$. Similarly, we found that $T_A = \sqrt{T_0^2 + T_{Ay}^2} = 87.36\rho g$ and $\theta_A = 57.17°$.

We can plot the cable shape, a catenary curve, based on Equations (11.25k) and (11.25n), and it is shown in (d), compared to a parabolic curve.

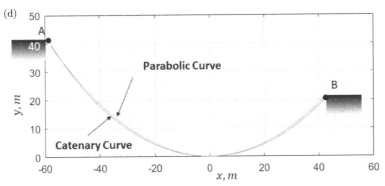

The curvature of the cable is not related to the density of the cable as (d) is plotted without its knowledge.

Finally, if we have a suspended bridge as shown in (e), the weight distribution over the cable is uniform along the x-axis. We modify Equations (11.25d), (11.25e), and (11.25g) as

$$T_x \sin\theta_x = \rho x g \qquad\qquad (11.25t)$$

$$\tan\theta_x = \rho x g / T_0 \qquad\qquad (11.25u)$$

$$\frac{dy}{dx} = \frac{\rho x g}{T_0} = \frac{x}{a} \qquad\qquad (11.25v)$$

From Equation (11.25g), we have $y = \frac{1}{2a}x^2$, which is a parabolic curve. The catenary curve and the parabolic curve are quite similar as shown in (d), but the settings are different.

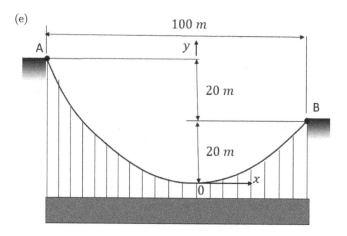

(e)

Problem 11.26: The uniform plank ABC has a mass of 100 kg and is supported at B by the large 150-kg cylinder and at A by the horizontal plane, as shown in (a). The coefficient of friction between the surfaces at A and B is 0.40 and that between the surfaces at D is 0.60. Calculate the force P required to start at end A moving to the left. Also compute the corresponding friction force acting at D. The small guiding roller at E turns with negligible friction.

(a)

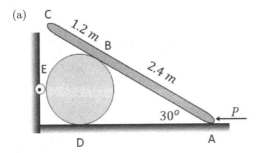

Solution: The solution, of course, starts with constructing proper FBDs. We first construct one for the combined system of the plank and the cylinder, as shown in (b).

(b)

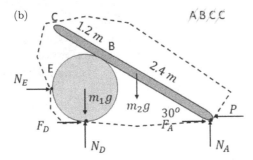

In (b), we are not sure about the direction of the frictional force, F_D. We simply assume one. However, there are at least six unknowns involved in (b). But we can only have five equations (three force equilibrium equations plus two friction force equations, if sliding happens at points A and D). Therefore, FBD of (b) is of no use.

We then construct two FBDs just for the cylinder and the plank, separately, as shown in (c) and (d), respectively. For (c), there are two contact points involving frictional forces. At the contact point B, both the plank and the cylinder will have moving tendencies; therefore, it is difficult to decide the direction of the frictional force at point B. However, if the plank is to move to the left, the likely rotation tendency of the cylinder is counter-clockwise. Therefore, we can be sure that the frictional force, F_D, is pointing to the left, different from our guess in (a). The FBD of (d) has five unknowns. We can assume sliding at points A and B. Therefore, we can have five equations. With closer examination, the force, N_B, can be easily solved with a moment equation with respect to point A.

We have

$$-N_B \cdot 2.4 + m_2 g \cos 30° \cdot 1.8 = 0 \tag{11.26a}$$

From which, we have $N_B = 637.18\,(N)$. Now, we have four more equations,

$$-P + F_A + F_B \cos 30° + N_B \sin 30° = 0 \tag{11.26b}$$

$$N_A - m_2 g + N_B \cos 30° - F_B \sin 30° = 0 \tag{11.26c}$$

$$F_A = 0.4\,N_A \tag{11.26d}$$

$$F_B = 0.4\,N_B \tag{11.26e}$$

From them, we have $N_A = 556.63\,(N)$, $P = 751.87\,(N)$, $F_A = 222.65\,(N)$, and $F_B = 254.87\,(N)$.

Now, from (c), with a moment equation with respect to the center of cylinder, we have $F_D = F_B = 254.87\,(N)$.

To determine other forces in (c), we need to construct corresponding equations. We already know that $F_D = F_B$. However, N_D must be much higher than N_B due to the weight of the cylinder. The sliding coefficient at point D is given as 0.6. As a result, $F_D \neq 0.6 N_D$. The only possibility is that the cylinder does not reach the threshold of rotation. The correct equation between N_D and F_D, should be

$$F_D = a N_D \tag{11.26f}$$

where a is an unknown friction coefficient value up to the limit of the static friction coefficient. As a result, we still have three unknowns in (c), and the other two equations are

$$N_D - m_1 g + F_B \sin 30° - N_B \cos 30° = 0 \tag{11.26g}$$

$$N_E - F_D - F_B \cos 30° - N_B \sin 30° = 0 \tag{11.26h}$$

Solving Equations (11.26f)–(11.26h), we obtain $N_E = 794.18\,(N)$, $N_D = 1895.87\,(N)$, and $a = 0.1$.

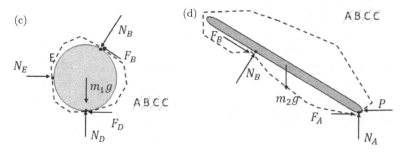

This is a seemingly easy problem, but unless one pays attention to the frictional force direction correctly and identifies that $F_D \neq 0.6 N_D$, the correct answers cannot be obtained.

With the above result, it becomes more clear how the plank will move due to P. It would slide over the cylinder while the cylinder stays stationary without rotation, held in place by frictional force F_D.

It would be interesting to work on the case if P is applied in the opposite direction, pulling the plank to the right. In this case, Equation (11.26a) is still valid and we have the same $= 637.18\,(N)$, assuming that there is no tendency of rotation of the plank.

Let us also assume that the cylinder would not rotate as in the previous case. Also, note that $N_E = 0$ when P is pulling to the right. From there, we can have a moment equilibrium equation with respect to point D,

$$-F_B R(\cos 30° + 1) + N_B R \sin 30° = 0 \qquad (11.26\text{i})$$

We obtain $F_B = 170.71\,(N)$. The corresponding friction coefficient at point B is $\mu_B = \frac{170.71}{637.18} = 0.27 < 0.4$. Therefore, the plank and the cylinder are not sliding with respect to each other.

$$-P - F_A - F_B \cos 30° + N_B \sin 30° = 0 \qquad (11.26\text{j})$$

$$N_A - m_2 g + N_B \cos 30° + F_B \sin 30° = 0 \qquad (11.26\text{k})$$

$$F_A = 0.4\,N_A \qquad (11.26\text{l})$$

Solving the above equations, we have $N_A = 342.83\,(N)$, $F_A = 137.13\,(N)$, and $P = -33.62\,(N)$. However, because we assume that P is pulling to the right, the results show that it is pushing to the left. As a result, the cylinder cannot stay stationary as in the first case. This problem now turns into a dynamic problem and we will need additional information in order to solve it. This is beyond the scope of this book.

Problem 11.27: The slender bar of length l and mass m is pivoted freely about point O as shown in (a). The spring has an unstretched length of $\frac{l}{2}$. Determine the equilibrium positions, excluding $\theta = \pi$, and determine the maximum value of the spring stiffness k so that the equilibrium is stable when $\theta = 0$.

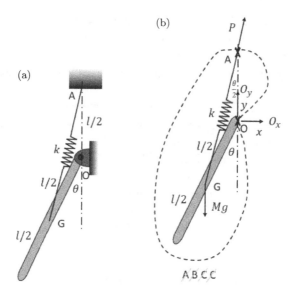

Solution: As in the case of Problem 11.22, we will invoke the method of virtual work to solve this problem. The first step is still to construct an FBD for the entire system as shown in (b).

From (b), we conclude that there are two non-conservative forces, O_x and O_y, but neither is doing work. Both P and mg are conservative forces. Therefore, we only need to define the corresponding potential energies. We have

$$U = U_k + U_g \tag{11.27a}$$

By specifying point O as the datum, we have

$$U_g = -mg\left(\frac{l}{2}\right)\cos\theta \tag{11.27b}$$

The elastic potential energy is

$$U_k = \frac{1}{2}k\left(l\cos\frac{\theta}{2} - \frac{1}{2}l\right)^2 \tag{11.27c}$$

Applying the method of virtual work, we have

$$\delta W - \delta U = -\delta U = -\left(\frac{1}{2}l\,mg\sin\theta - \frac{1}{2}kl\left(l\cos\frac{\theta}{2} - \frac{1}{2}l\right)\sin\frac{\theta}{2}\right)(\delta\theta) = 0 \tag{11.27d}$$

Substituting $\sin\theta = 2\sin\frac{\theta}{2}\cos\frac{\theta}{2}$ into Equation (11.22d), we have

$$l\,mg\sin\frac{\theta}{2}\cos\frac{\theta}{2} - \frac{1}{2}kl\left(l\cos\frac{\theta}{2} - \frac{1}{2}l\right)\sin\frac{\theta}{2} = 0 \qquad (11.27\text{e})$$

The solution to Equation (11.22e) is either

$$\sin\frac{\theta}{2} = 0 \qquad (11.27\text{f})$$

or

$$mg\cos\frac{\theta}{2} - \frac{1}{2}k\left(l\cos\frac{\theta}{2} - \frac{1}{2}l\right) = 0 \qquad (11.27\text{g})$$

The two possible equilibrium positions are found to be

$$\theta = 0 \quad \text{or} \quad \theta = 2\cos^{-1}\left(\frac{1}{2 - 4\frac{mg}{kl}}\right)$$

Now, let's examine the stability at $=0$. We will determine $\frac{\delta^2 U}{\delta U^2}$, which is

$$\frac{\delta^2 U}{\delta\theta^2} = \frac{1}{2}mgl\cos\theta - \frac{1}{4}kl\left(l\cos\frac{\theta}{2} - \frac{1}{2}l\right)\cos\frac{\theta}{2} + \frac{1}{4}kl^2\sin^2\frac{\theta}{2} \qquad (11.27\text{h})$$

Introducing $\theta = 0$ into Equation (11.27h), we have

$$\left.\frac{\delta^2 U}{\delta\theta^2}\right|_{\theta=0} = \frac{1}{2}mgl - \frac{1}{8}kl^2 \qquad (11.27\text{i})$$

To ensure stability, we must have $\left.\frac{\delta^2 U}{\delta\theta^2}\right|_{\theta=0} > 0$, which leads to $k < \frac{4mg}{l}$.

Problem 11.28: Determine the equilibrium position of the mechanism shown in (a). The rod BC is held through a swivel block A. As the parallel beams swing up, rod BC will pass through swivel A to stretch a spring. The spring is upstretched when $\theta = 0$. A moment $M = 60\,\text{Nm}$ is applied to swing the beams upward. The spring has a stiffness $k = 165\,\text{N/m}$. The two parallel beams have a mass $m_1 = 7.5\,\text{kg}$ each, and the horizontal beam has a mass $m_2 = 10\,\text{kg}$.

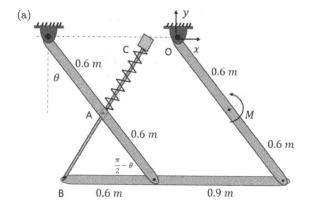

Solution: Just as in Problem 11.27, we quickly determine that there are
no non-conservative forces doing any work except the applied moment M.
Invoking the method of virtual work, we have

$$\delta W - \delta U = 0 \tag{11.28a}$$

where $\delta W = M\delta\theta$. By defining a datum at point O, the potential energy
can be found to be

$$U = U_k + U_g \tag{11.28b}$$

where

$$U_g = 2 \times (-0.6\, m_1 g \cos\theta) + (-1.2\, m_2 g \cos\theta) \tag{11.28c}$$

The elastic potential energy of the spring needs to be determined by finding
out the deformation of the spring. From (a) and (b), the spring stretch at
an angle θ is found to be

$$\Delta l = 0.6\sqrt{2} - 2 \times 0.6 \sin\left(\frac{\pi}{4} - \frac{\theta}{2}\right) \tag{11.28d}$$

The corresponding elastic potential energy is

$$U_k = \frac{1}{2}k\left(0.6\sqrt{2} - 2 \times 0.6 \sin\left(\frac{\pi}{4} - \frac{\theta}{2}\right)\right)^2 \tag{11.28e}$$

From the above equations, Equation (11.28a) becomes

$$M - k\left(0.6\sqrt{2} - 2 \times 0.6\sin\left(\frac{\pi}{4} - \frac{\theta}{2}\right)\right)\left(1.2\cos\left(\frac{\pi}{4} - \frac{\theta}{2}\right)\right)\left(\frac{1}{2}\right)$$
$$- 1.2 m_1 g \sin\theta - 1.2 m_2 g \sin\theta = 0 \qquad (11.28f)$$

Solving Equation (11.28f), we have $\theta = 14.38°$. Of course, Equation (11.28f) is not easy to solve by hand. We can use Matlab to solve it as in Problem 11.25. However, the Matlab command "solve" does not provide correct answers because it uses symbolic solutions. Instead, using Matlab command "vpasolve" to solve the equation numerically, one will get the result quickly. Here is the script:

```
clear th
syms th
format long
% th is the angle ;

g=9.810;
m1=7.5;
m2=10;
k=165;
M=60;

eq1 =M-k*(0.6*2^0.5 - 1.2*sin(pi/4-th/2))*(0.6*cos(pi/4-th/2))
-1.2*(m1+m2)*g*sin(th) ==0;

eqns = [eq1 ];

S = vpasolve(eqns)
sol = [S]
```

We can also plot the right-hand side of Equation (11.28f) with respect to θ, as shown in (c). We will find the same answer when the value becomes zero. Also from (c), the slope at the equilibrium is negative, indicating the equilibrium is unstable. Therefore, if the applied moment is lost, the equilibrium cannot be maintained. This is similar to the flying of an airplane. If an airplane loses its power, it will fall from sky.

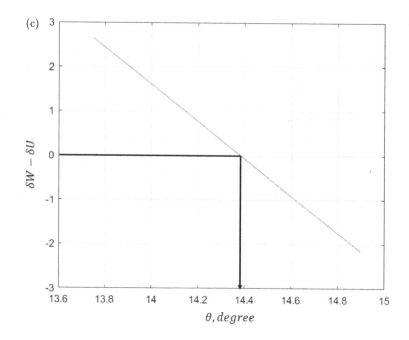

Problem 11.29: The circular disk with a central hole is subjected to a uniform compressive stress around its rim, as shown in (a). The thickness t of the disk is a function of r only but is small everywhere compared with r. Derive the differential equation of equilibrium in the radial direction for a differential element of the disk.

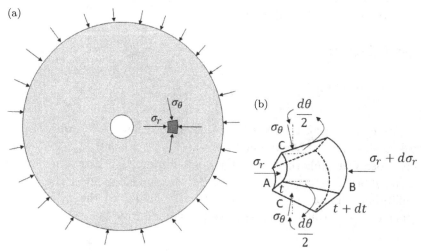

Solution: First, we will enlarge a 3D differential element as shown in (b). This differential element is a free body. The boundary is not drawn here. This element is constrained at four sides (not top and the bottom sides) by distributed forces (stresses). Now we only have to construct the equilibrium force equation along the r direction.

The four sides are marked as A, B, and C, as shown in (b). The force acting at surface A is $\sigma_r A_A = \sigma_r r d\theta\, t$, where A_A is the area of surface A. The force acting at surface B is $-(\sigma_r + d\sigma_r)A_B = -(\sigma_r + d\sigma_r)(r + dr)\, d\theta(t + dr)$. The force acting on each surface C is $\sigma_\theta A_C = \sigma_\theta\, dr\, t$. The component in the r direction is $\sigma_\theta\, dr\, t\frac{d\theta}{2} \times 2$. Putting all these forces together, we have

$$\sigma_r r d\theta\, t - (\sigma_r + d\sigma_r)(r + dr)d\theta(t + dr) + \sigma_\theta\, dr\, t\, d\theta = 0 \qquad (11.29a)$$

Cleaning up Equation (11.29a), we have

$$\frac{\sigma_r - \sigma_\theta}{r} + \frac{\sigma_r}{t}\frac{dt}{dr} + \frac{d\sigma_r}{dr} = 0 \qquad (11.29b)$$

Problem 11.30: The 500 kg trailer with center of mass at G is being towed at a constant speed up a slope, as shown in (a). The right wheel suddenly locks. If the trailer maintains the same speed even with one wheel dragging, calculate the total force acting at the ball joint A. The sliding friction coefficient between the wheel and the ground is 0.8 and $\theta = \tan^{-1} 0.1$.

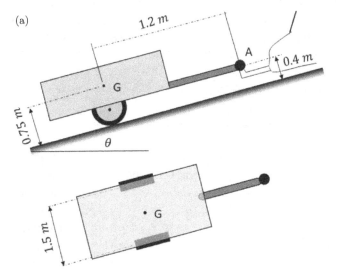

(a)

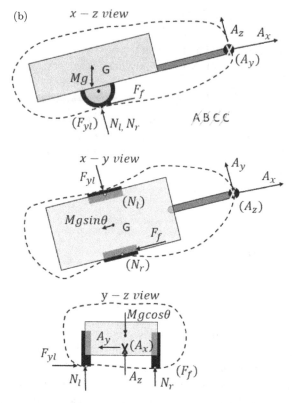

Solution: As usual, we need to construct a proper FBD. This is a 3D problem, and we will construct the FBDs in three views, as shown in (b). When constructing 2D views for a 3D problem, we should draw all three views out first. When we go through the ABCC method, we will mark on all three views. If a force is perpendicular to the viewing plane, we still mark the force but put it in parentheses because it is not visible in that particular view.

We will isolate the trailer at ball joint A. There is no applied force (A), there is a body force (B), there are two contact points (C) at the tires on the ground, and there is a constraint point (C) at the ball joint. To mark the body force, we draw Mg at point G on the $x-z$ view. On the $x-y$ view, we actually see $Mg\sin\theta$, and on the $y-z$ view we actually see $Mg\cos\theta$. They are marked accordingly in (b).

For the contact points at the tires, we see both normal forces, N_l and N_r, on the left and right tires, respectively, in the $x-z$ and $y-z$ views. In the $x-y$ view, these two normal forces are "invisible"; therefore, we label them in parentheses. For the frictional forces, it would take some consideration. Because the right tire is locked, it would be sliding over the ground. We know that $F_f = \mu_k N_r$. Because the total frictional force of the right tire cannot exceed $\mu_k N_r$, we know that F_f is the only frictional force acting on the right tire. For the left tire, because it is still rolling, we will assume that the rolling friction is zero. However, if we consider the moment equilibrium with respect to ball joint A, the frictional force, F_f, will produce a moment which must be countered by a lateral force on the left tire. If the trailer does not slide in the lateral direction (y-direction), then this lateral force is a frictional force within the static friction limit. We denote this lateral force on the left tire as F_{yl}, which is visible in the $x - y$ and $y - z$ views, while invisible in the $x - z$ view.

For the constraint force at ball joint A, we will have three forces, A_x, A_y, and A_z. We mark them accordingly in all three views. This completes the ABCC procedure to construct a 3D FBD systematically.

There are seven unknowns involved in (b). They are N_l, N_r, F_f, F_{yl}, A_x, A_y, and A_z. We will need seven equations. We do have six force/moment equilibrium equations plus the friction equation of the right tire. These seven equations are

$$\left(\sum \vec{F}\right)_x = 0 \Rightarrow A_x - F_f - mg\sin\theta = 0 \tag{11.30a}$$

$$\left(\sum \vec{F}\right)_y = 0 \Rightarrow A_y - F_{yl} = 0 \tag{11.30b}$$

$$\left(\sum \vec{F}\right)_z = 0 \Rightarrow N_l + N_r - mg\cos\theta + A_z = 0 \tag{11.30c}$$

$$\left(\sum \vec{M}\right)_{G,x} = 0 \Rightarrow -0.35\,A_y + 0.75\,F_{yl} - 0.75\,N_l + 0.75\,N_r = 0 \tag{11.30d}$$

$$\left(\sum \vec{M}\right)_{G,y} = 0 \Rightarrow -0.75\,F_f + 1.2\,A_z + 0.35\,A_x = 0 \tag{11.30e}$$

$$\left(\sum \vec{M}\right)_{G,z} = 0 \Rightarrow -0.75\,F_f + 1.2\,A_y = 0 \tag{11.30f}$$

$$F_f = 0.8\,N_r \tag{11.30g}$$

Note that the moment equilibrium equations with respect to the mass center, G, can be constructed using corresponding views. For example, for $(\sum \vec{M})_{G,x} = 0$, we should use the $y - z$ view.

Again, we can solve these seven equations either by hand or by using Matlab. We obtain $N_l = 2551.5\,(N)$, $N_r = 1982.8\,(N)$, $F_f = 1586.2\,(N)$, $F_{yl} = 991.4\,(N)$, $A_x = 2074.3\,(N)$, $A_y = 991.4\,(N)$, and $A_z = 386.4\,(N)$.

The total force acting at ball joint A is

$$A = \sqrt{A_x^2 + A_y^2 + A_z^2} = 2331.3\,(N) \qquad\qquad (11.30\text{h})$$

Let us verify the friction coefficient at the right tire, $\mu_r = \frac{F_f}{N_r} = 0.8$. At the left tire, $\mu_l = \frac{F_{yl}}{N_l} = 0.39 < 0.8$, which satisfies the assumption of the not sliding condition at the left tire. Finally, note that the normal forces at the left and right tires are not equal. This is due to the lateral weight shift caused by F_f. This lateral weight shift could happen when a car is making a turn.

Problem 11.31: A mass m slides with negligible friction along the smooth rod mounted in the frame, which can pivot freely about a horizontal axis through O, as shown in (a). When the frame is horizontal, the mass is at the center and the springs are neither stretched nor compressed. Each spring has a spring constant of $\frac{k}{2}$. The frame has a negligible mass. Determine the equilibrium positions defined by the angle of rotation of the frame. Investigate the stability of these equilibrium positions.

(a)

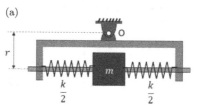

Solution: Let us examine the force equilibrium at an equilibrium position when $\theta \neq 0$, which is shown by the FBD of (b). Because there are only two forces involved in (b), these two forces must be the same size and in the opposite directions, as shown in (b). From this observation, we can define the displacement of the mass along the rod, x, as a function of θ. We have

(b)

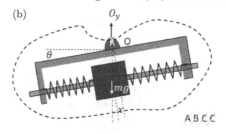

ABCC

$$\frac{x}{r} = \tan\theta \tag{11.31a}$$

The vertical distance of the center of the mass with respect to point O is then found to be

$$y = \sqrt{x^2 + r^2} \tag{11.31b}$$

From the above two equations, we also have

$$\frac{x}{y} = \frac{x}{\sqrt{x^2 + r^2}} = \sin\theta \tag{11.31c}$$

For this system, only the conservative, gravitational, and spring forces are doing work. Invoking the method of virtual work, we have

$$-\delta U = 0 \tag{11.31d}$$

By defining a datum at point O, the potential energy can be found to be

$$U = U_k + U_g \tag{11.31e}$$

where

$$U_g = -mgy = -mg\sqrt{x^2 + r^2} \tag{11.31f}$$

The elastic potential energy of the spring is determined by the deformation of the spring, which is

$$U_k = 2 \times \left(\frac{1}{2}\frac{k}{2}(x)^2\right) = \frac{1}{2}kx^2 \tag{11.31g}$$

From the above equations, Equation (11.31d) becomes

$$\frac{mgx}{\sqrt{x^2 + r^2}} - kx = 0 \tag{11.31h}$$

Introducing Equations (11.31a) and (11.31c) to (11.31h), we have

$$-mg\sin\theta + kr\tan\theta = 0 \tag{11.31i}$$

We then obtain the possible equilibrium angle as

$$\bar{\theta} = \cos^{-1}\frac{kr}{mg} \tag{11.31j}$$

Because $\cos\theta = r/\sqrt{x^2 + r^2}$, the corresponding displacement of the mass along the rod is found to be

$$x = \frac{mg}{k}\sqrt{1 - (kr/mg)^2} \tag{11.31k}$$

To evaluate the stability, we check the value of

$$\frac{\delta^2 U}{\delta\theta^2}\bigg|_{\bar\theta} = -mg\cos\theta + kr/(\cos\theta)^2 = -kr + (mg)^2/(kr) \qquad (11.311)$$

To ensure $\frac{\delta^2 U}{\delta\theta^2}|_{\bar\theta} > 0$, we need to have $k < \frac{mg}{r}$ to render the equilibrium stable at $\bar\theta = \cos^{-1}\frac{kr}{mg}$.

In fact, $\theta = 0$ is also an equilibrium position. The value of $\frac{\delta^2 U}{\delta\theta^2}|_{\theta=0} = -mg + kr$. If $k < \frac{mg}{r}$, then $\theta = 0$ is an unstable equilibrium. Any tilting of the frame will rotate and stabilize at $\bar\theta$. On the other hand, if we make $k > \frac{mg}{r}$, $\theta = 0$ is stable; thus, the frame will stay horizontal.

Index